Energy-Efficient Approaches in Industrial Applications

Edited by Murat Eyvaz, Abdülkerim Gok and Ebubekir Yüksel

Published in London, United Kingdom

IntechOpen

Supporting open minds since 2005

Energy-Efficient Approaches in Industrial Applications
http://dx.doi.org/10.5772/intechopen.74268
Edited by Murat Eyvaz, Abdülkerim Gok and Ebubekir Yüksel

Contributors
Xiufang Ye, Dongchu Chen, Ibrahim Udale Hussaini, Sandra G. L. Persiani, Alessandra Battisti, Tran Anh

Notice
Statements and opinions expressed in the chapters are these of the individual contributors and not
necessarily those of the editors or publisher. No responsibility is accepted for the accuracy of
information contained in the published chapters. The publisher assumes no responsibility for any
damage or injury to persons or property arising out of the use of any materials, instructions, methods
or ideas contained in the book.

First published in London, United Kingdom, 2019 by IntechOpen
IntechOpen is the global imprint of INTECHOPEN LIMITED, registered in England and Wales,
registration number: 11086078, The Shard, 25th floor, 32 London Bridge Street
London, SE19SG – United Kingdom
Printed in Croatia

British Library Cataloguing-in-Publication Data
A catalogue record for this book is available from the British Library

Additional hard copies can be obtained from orders@intechopen.com

Energy-Efficient Approaches in Industrial Applications
Edited by Murat Eyvaz, Abdülkerim Gok and Ebubekir Yüksel
p. cm.
Print ISBN 978-1-78985-519-7
Online ISBN 978-1-78985-520-3

We are IntechOpen,
the world's leading publisher of
Open Access books
Built by scientists, for scientists

4,000+
Open access books available

116,000+
International authors and editors

120M+
Downloads

Our authors are among the

151
Countries delivered to

Top 1%
most cited scientists

12.2%
Contributors from top 500 universities

Interested in publishing with us?
Contact book.department@intechopen.com

Numbers displayed above are based on latest data collected.
For more information visit www.Intechopen.com

Meet the editors

Dr. Murat Eyvaz is an assistant professor of the Environmental Engineering Department at the Gebze Technical University. He received his BSc. degree in Environmental Engineering from Kocaeli University in Turkey in 2004. He completed his M.Sc. and Ph.D., 2013 in Gebze Institute of Technology (former name of GTU) in Environmental Engineering. He completed his post-doctoral research at the National Research Center on membrane technologies in 2015. His research interests are water and wastewater treatment, electrochemical processes, filtration systems/membrane processes, and spectrophotometric and chromatographic analyses. He has co-authored numerous journal articles and conference papers and has taken part in many national projects. He serves as an editor in 30 journals and a reviewer in 100 journals indexed in SCI, SCI-E, and other indexes.

Dr. Abdülkerim Gök is a research associate at the Gebze Technical University, Gebze, Kocaeli, Turkey. He completed his bachelor degree in Materials Science and Engineering at Anadolu University, Turkey in 2007. He received his M.Sc. degree in Chemical Engineering from Columbia University, New York, USA, in 2011 and his PhD. degree in Materials Science and Engineering from Case Western Reserve University, Cleveland, OH, USA, in 2016. His work focuses on developing predictive and mechanistic degradation pathway models of polymeric materials used in photovoltaic module materials under accelerated and natural real-world weathering exposures. His research interests include lifetime and degradation science of PV module and module materials, reproducible statistical methods, and the effect of environmental stressors on PV module performance.

Prof. Ebubekir Yüksel is a faculty member of the Environmental Engineering Department at the Gebze Technical University. He received his bachelor degree in Civil Engineering from İstanbul Technical University in 1992. He completed his graduate work (M.Sc., 1995 and Ph.D., 2001) at the İstanbul Technical University. His research interests are applications in water and wastewater treatment facilities, electrochemical treatment process and filtration systems at the laboratory and pilot scale, watershed management, flood control, deep sea discharges, membrane processes, spectrophotometric analyses, chromatographic analyses, and geographic information systems. He has co-authored numerous journal articles and conference papers and has taken part in many national projects. He has produced more than 25 peer-reviewed publications in journals indexed in SCI, SCI-E, and other indexes.

Contents

Preface

A large amount of energy is consumed in the industry to meet the power needed for production processes. In order to meet the heat and mechanical power needs required for many industrial processes, natural gas, petroleum fuel, and electricity are mostly used as energy sources. In addition to the efficient use of energy in order to reduce operating costs in industrial applications, alternatives such as efficient use of energy for conservation of resources and climate, energy recovery, renewable energy preferences, and energy production from wastes are becoming more common. With proper energy management, it is possible to increase energy efficiency independently of the size of the industry and the technologies used in the process. The development of new alternatives for energy efficiency and saving is crucial to meet the growing world energy needs and to compete effectively with fossil fuels and thus reduce greenhouse gases.

This book, which consists of four chapters, includes studies on energy efficiency. In the first chapter, the properties and types of thermal insulation coatings used for energy saving are reported. In the second chapter, research including the artificial neural network method used to increase energy efficiency by reducing the fuel consumption of a ship is proposed. In the third chapter, a strategic plan for energy efficiency implementations by the government and housing stakeholders towards the housing sector is proposed. In the last chapter, the latest examples of membrane structures, and materials used in the context of experimental design, art, and architecture and their relationship with the environment and energy are reviewed.

Asst. Prof. Murat Eyvaz and Prof. Ebubekir Yüksel
Department of Environmental Engineering,
Gebze Technical University, Turkey

Dr. Abdülkerim Gök
Department of Materials Science and Engineering,
Gebze Technical University, Turkey

Chapter 1

Thermal Insulation Coatings in Energy Saving

Xiufang Ye and Dongchu Chen

Abstract

The surface temperature of object rises due to the accumulated heat when it absorbs solar energy, the excessive temperature caused by solar radiation will result in many inconveniences and even troubles in industrial production and daily life; in order to maintain the proper temperature of the object, a large amount of energy is consumed. The development of effective and economic thermal insulation materials is the key to meet the urgent needs for energy saving and emission reduction. In the face of variety of choices of thermal insulation materials, thermal insulation coating become more and more popular due to its good thermal insulation performance, economic, easy to use, and adaptability for a wide range of substrates. With the thermal insulation functional fillers (briefly called fillers in the following text) in coating system, the films can show a certain thermal insulation effect by reflecting, radiating, or isolating heat. As a result, when covered by thermal insulation coatings, the surface temperature of object would be greatly decreased. In this case, a large amount of energy consumed for cooling down the objects exposed to sunlight could be saved, which means the energy consumption can be reduced effectively by just covering with a thermal insulation coating on the surface of object.

Keywords: thermal insulation coatings, mechanism, composite, functional fillers, energy saving

1. Introduction

Under the sunlight, the object absorbs solar energy and the surface temperature rises. As we all know, when there is a temperature difference between the surface and the interior of the object, the heat transfer occurs. As a result, the temperature inside the object also increases by heat transfer. Just like the room temperature become warmer in winter or become hotter in summer after sunrise, and on the other hand, the room temperature become colder in winter or become cooler in summer after sunset. In order to maintain the suitable room temperature, air conditioners are widely used; however, it consumes too much energy. Relevant statistics [1, 2] show that more than 50% of human material obtained from nature is used to build various types of buildings and their ancillary facilities, and at least 50% of energy in the world was consumed during the construction and for the use of these buildings.

In addition, the higher surface temperature caused by solar radiation can affect both industrial production and daily life. As we all know, high temperature is disadvantageous to store food, vegetables, fruits, and medicines. So, a very large amount of energy has been consumed for sprinklers, air conditioners, and fans. But moreover, high temperature accelerates the corrosion, aging, and degradation rate of materials, so that these materials will be limited in application due to its affected mechanical and chemical properties. For example, higher temperature will cause the thermal expansion and thermal stress of materials, which will accelerate cracking, corrosion, and destruction of the material. In some cases, high temperatures cause not only much more energy consumption but also the potential hazards. If we take petrochemical containers, for example, in hot summer, the storage tank for oil, gas, chemical, etc. required to be cooled by water spray regularly. Otherwise, high temperatures can cause excessive volatile organic compounds volatilization and even explosion. But this cooling method not only wastes a lot of water and electricity costs, but also affects the tank equipment maintenance.

As discussed above, large amounts of energy have already been consumed and are still consuming to adjust the proper temperature in industrial production and daily life. Among them, the proportion of building energy consumption is still the largest one. Moreover, the building energy consumption is now increasing rapidly with the increasing building scale [3]. So, the development of building energy-saving technologies has become an urgent need for all countries in the world. In the existing building energy-saving technologies, the choice of the plan of the external envelope is the first issue to be considered in the building energy-saving design. The thermal performance of the external envelope is the basis for determining whether the building can save energy [2, 4–6]. Energy conservation of the external envelope mainly from the following aspects, including the building's walls, roofs, doors, and windows, in addition to the rationality of building structures, proceeds on the walls [7], roofs [5], doors, and windows [6]; the most crucial part is the use of thermal insulation material to achieve the interval thermal insulation effect and to achieve the purpose of building energy efficiency.

In the face of variety of choices of thermal insulation materials, thermal insulation coating become more and more popular for its economic, easy to use, suitable for variable substrates, and good thermal insulation effects. Generally speaking, coatings are basically composed of resin, functional fillers, additives, pigments, and solvents. Compared with other coatings, the most significant characteristic of thermal insulation coatings is that the functional fillers used are materials with excellent thermal insulation performance. Usually, these fillers are called thermal insulation functional fillers (shortly referred as fillers in the following text). There are already plenty of coating products composed of different resins that are suitable for different substrates. So in theory, it is possible to get various thermal insulation coatings that are suitable for walls, roofs, doors, and even windows just with the combination of different resins and fillers. From this point of view, suitable fillers are chosen that are crucial to achieve a desired thermal insulation performance.

2. The mechanism of thermal insulation coatings

According to the heat transfer theory, the solar radiation is mainly transmitted to the object as heat, so when the surface of the object absorbs sunlight, the heat can transfer from the surface to the inside of object. As a result, the temperature of the object rises accordingly. But if the object can be covered with thermal insulation coatings on its surface, most of the extra heat from sunlight can be insulated before it transfers to the surface of the object (**Figure 1**).

The heat transfer is usually a combination of heat conduction, heat convection, and heat radiation. Based on these, there are three different thermal insulation modus: obstructive, reflective, and radiative. Accordingly, with corresponding fillers, thermal insulation coatings can be divided into four different kinds: obstructive, reflective, radiative, and composite thermal insulation coatings [8].

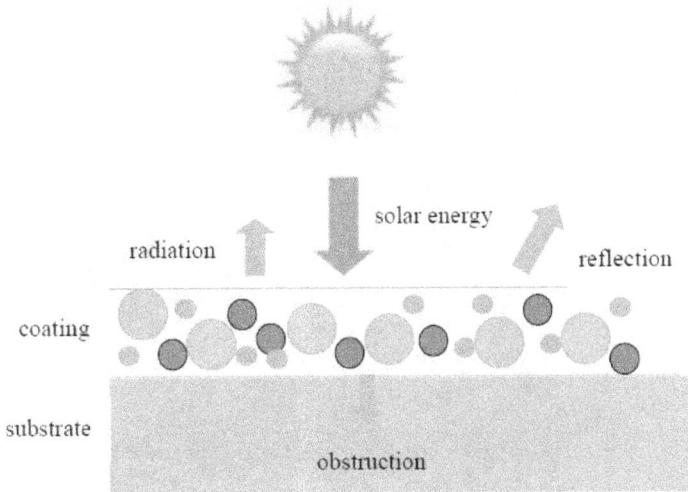

Figure 1.
Schematic diagram of the mechanism of thermal insulation with coating.

2.1 Obstructive thermal insulation coatings

The obstructive thermal insulation coating is a kind of passive thermal insulation coating by resisting heat transfer with particular fillers. But as mentioned above, the coatings system always consists of resin, fillers and pigments, additives, and solvents. So, not only the thermal conductivity of fillers but also other materials are crucial to the heat-resist performance of film. In general, pigments, fillers, additives, and film-forming materials with low thermal conductivity are selected to produce an obstructive thermal insulation coating; among them, the fillers with very low thermal conductivity called thermal insulation functional fillers are the key to achieve an excellent thermal insulation performance of the film.

With these special fillers, the film can stay at a low thermal conductivity and achieve an excellent thermal resistance performance. So, the thermal conductivity (λ) of the functional fillers is generally less than 0.06 W·m^{-1}·K^{-1} as the λ of air is about 0.0267 W·m^{-1}·K^{-1}, which means quite poor thermal conductivity, so most of the obstructive thermal insulation fillers have a hollow structure. Common thermal insulation functional fillers are materials with a hollow structure, such as, inorganic silicate-based materials, asbestos fibers, expanded perlite, sepiolite, closed-cell perlite, diatomaceous earth, and so on. Closely packed hollow particles in these fillers can form a layer of gas that has a barrier to heat and blocks the "thermal bridge" (**Figure 2**).

In practical applications, the film thickness always affects its thermal insulation effect. Generally speaking, thicker one means lower thermal conductivity and shows better heat insulate performance of the film. As a result, the coating is expected to be as thick as possible, based on these, the thickness of dry film is usually controlled at 5–20 mm for many obstructive thermal insulation coatings since 1980s [9, 10]. Although thicker films are needed to achieve better thermal insulation performance, but unfortunately, thicker films show the following problems at the same time: weaker impact resistance, obvious dry shrinkage, and high moisture absorption rate.

The situation will be quite different if the hollow structure of the fillers is closed, like hollow ceramic beads, hollow glass beads, hollow porous silica ceramics, etc. Studies [11–15] show that films with closed hollow structure have an excellent thermal insulation performance especially when the size of thermal insulation functional fillers reaches nanoscale, and it can be even used as a thin film. This

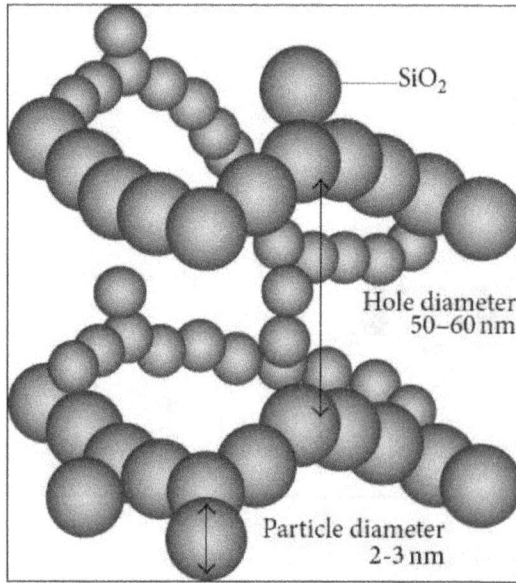

Figure 2.
Network architecture of an aerogel.

could be caused by the very low-close to zero-heat convection and heat conduction from molecular vibration when the bulk density of the coating and the pore diameter therein are sufficiently small. Reports shown unlike traditional thick-coated obstructive coatings, thin films with closed hollow structure fillers like hollow ceramic microbubbles can effectively enhance the thermal insulation performance of the buildings. A thin thermal insulation coating with silica hollow spheres as a functional filler is prepared, and the thermal insulation performance of the film was quantitatively evaluated by thermoresistance superposition method; test results showed that the thermal conductivity of the film is just 0.05 W·m^{-1}·K^{-1}, which means having an excellent thermal insulation effect [12]. Nano-TiO_2-modified hollow polymer microspheres were used as a functional filler in thermal insulation coating [16]; test data showed that the thermal conductivity of the film is only 0.1687 W·m^{-1}·K^{-1} and temperature difference of the film was up to 5.8°C.

Anyway, although a lot of works have been done and achieved a great progress in the improvement of thermal insulation performance in thin obstructive coatings, the main products on trading market are still traditional thick-coated obstructive coatings. As discussed above, the thick-coated obstructive coatings are not very suitable for buildings due to the contradiction between thermal insulation and comprehensive properties. So, more works are still needed to explore thin obstructive coating products in building energy saving.

2.2 Reflective thermal insulation coatings

The film of reflective thermal insulation coatings can reflect solar energy, rather than absorption or resist. Usually, we can use total solar reflectance (TSR) to evaluate the reflectivity of a material. TSR means the ratio of solar energy reflected by a certain surface of material, usually expressed as a percentage. For example, when the TSR value of a particular material is 75%, which means, the material can reflect away 75% of solar energy and only absorb the rest 25% of solar energy. Theoretically, any material can reflect solar energy more or less. As the energy wavelength of solar radiation is mainly concentrated at the range of 200–2500 nm, to specific, about 50% is distributed in the visible spectrum (from 400 to 720 nm) and 43% distributed in the near-infrared spectrum (from 720 to 2500 nm). Since higher reflectivity means better thermal insulation of the film in 400–2500 nm region, the first principle to choose reflective fillers is that the material should show high reflectivity in visible and near-infrared spectrum. Researches [17, 18] show that these fillers can improve the thermal insulation performance of the film obviously compared with the traditional thermal insulation materials.

Usually, the visible color of the film is decided by the visible color of the fillers, and the fillers show particular color due to its selective reflection and absorption of visible spectrum. For example, white means the filler almost completely reflects all visible spectrum from 400 to 720 nm, whereas black means the filler absorbs almost all visible spectrum and red means the filler can primarily reflect spectrum from 650 to 700 nm, while absorb other spectrums in visible region. Based on these, white is the best color for infrared reflective fillers, for white fillers can reflect away almost all the spectrum in visible bands. For example, the TSR of titanium white is higher than 75%. But on the opposite, black fillers are barely selected for thermal insulation, because it can absorb almost all solar energy in visible spectrum, such as the TSR of carbon black is as low as 3–5%, which means it can absorb 95–97% of the solar energy. How different colors affect the indoor temperature of building has been studied not only theoretically but also experimentally [19, 20–23]; test in different conditions verified that the white fillers show better thermal insulation than fillers with other colors, especially black. Just after hours of solar radiation, the room temperature is 7°C higher when the surface of building is covered with black than white. Taking decorative into account, pigments show particular colors are usually added to coatings; so eventually, the reflection and absorption properties of film in the visible spectrum are affected by both fillers and pigments.

As discussed above, adding reflective fillers into coatings is an effective way to improve thermal insulation performance of the film. In this situation, fillers with high reflectivity at both visible spectrum and near-infrared spectrum bands are good choice. Under this premise, metal, metal oxide, hollow glass beads, fly ash beads and ceramic beads, and other materials with higher reflectivity are mainly selected as functional fillers when reflective thermal insulation coatings are prepared. But it is worth noting that, not only the color but also the structure of fillers affects the thermal insulation performance of the film; for example, metal oxide fillers with nanocrystal structure have better near-infrared reflectivity, which means better thermal insulation performance than with ordinary structure of metal oxide fillers [24–26].

Sometimes, two or more kinds of reflective thermal insulation functional fillers are mixed in order to get better thermal insulation performance of the film. But the truth is that the reflectivity of the mixed fillers is not simply a sum of the reflectivity of each filler. For example, the TSR of CoAl blue and MnSbTi brown is 35.7 and 32.6%, respectively, but if these two fillers were mixed according to a mass ratio 1:1, test results show that the TSR value of the mixture is only 26.9%, which is not only lower than the intermediate value 34.15% but also lower than the minimum TSR value of MnSbTi brown (32.6%).

To reflective thermal insulation coating systems, as the reflection occurs mainly on the surface of the film, thicker films do not always mean better thermal insulation performance. This is quite different to obstructive thermal insulation coatings. Generally speaking, there is an optimal value thickness of the reflective thermal insulation film; if the thickness of the film is lower than the optimal value, the thermal insulation performance is better when the film is thicker, but if the thickness of the film exceeds the optimal value, increasing the thickness of the film shows little effect on improving the thermal reflection efficiency of the film. This is because when the film is thinner, part of the solar can penetrate the film and be absorbed by the substrate under the film, but when the thickness reaches to a certain value, the substrate is completely covered by the film and the reflectivity is stable at same time; as a result, the thermal insulation effect become steady [27, 28].

With reflective thermal insulation functional fillers, films can reflect solar directly back to atmosphere, rather than first absorb and the emission as the thermal conductive coating; so theoretically, the thermal insulation performance of reflective thermal insulation coatings is better than obstructive thermal insulation coatings [18]. It is noteworthy that reducing the roughness of the film surface is conducive to improving the thermal reflectivity of the film. Hollow glass microspheres covered with nickel were used as fillers [29]; the results showed that the thermal insulation performance of the film is excellent. ZrO_2 ceramic balls coated with potassium silicate have higher light scattering, reflectance, about 10–20 times that of common ZrO_2 ceramic balls. Compared with the same size of rutile TiO_2 fillers, the effect of modified ZrO_2 ceramic ball is improved by 1/3 [30].

Fillers with high reflectivity and high emissivity were applied to improve the reflectivity of the films in the near-infrared region (720–2500 nm) and visible region (400–720 nm). Researchers have done much and made big improvements in this area up to now; as a result, reflective thermal insulation coatings have already been studied and used widely [18, 31–33]. For example, covered with heat-reflective insulation film on exterior walls of building in Hangzhou, China, a typical hot summer and cold winter zone, the surface temperature of the wall can be reduced up to 10°C. By calculating, it was found that the annual air-conditioning electricity saving with heat reflective insulation coating on exterior walls is about 5.8 kWh/(m² month), which indicated that the energy saving effect with the heat insulation coating is obvious [34].

With good thermal insulation performance, various reflective thermal insulation coating products can be selected in coating markets, which is now the main product in thermal insulation coating market.

2.3 Radiative thermal insulation coatings

Any object exposed to the sunlight can absorb while radiate solar energy at the same time. If the object absorbs more energy from solar than it radiates to the external space, the temperature of the object increases. On the other hand, if the object radiates more energy than it absorbs, the temperature of the object decreases. During this progress, the radiated energy is emitted in the form of invisible infrared light and longer wavelength electromagnetic waves. This radiation caused by molecular, atomic thermal motion is called thermal radiation.

Theoretically, thermal radiation exists between any practicality object. That means when any object radiates the energy of itself into external space, the external space radiates energy back to the object at the same time. Although the two processes always exist at the same time, but as we all know, when the temperature of the object is higher than the external space, the results of thermal radiation are that the object transmits more energy to external space and vice versa. If the temperature between object to external space is the same, there is no temperature change for the object after thermal radiation, for the amount of energy transmit, and accept by object is equal during the whole process. The temperature in outer space is close to absolute 0 K, so it seems that outer space is an ideal energy receptor, which means that any object can radiate the energy of itself into outer space with thermal radiation. But unfortunately, the energy radiation from objects on the ground to outer space is always been impeded by the outer surface atmosphere of the earth.

As the atmosphere worked as a barrier between the object and the outer space, so in order to get an ideal thermal radiation, first of all, we make sure the radiation can be successfully transmitted through the atmosphere into outer space. Atmosphere is mainly composed of water vapor and carbon dioxide, and these two substances show a weak absorbance during 8–13 μm spectrum. That is to say, the atmosphere has a high transmission during 8–13 μm radiation, or in other words, when the thermal radiation between object and outer space occurs during 8–13 μm spectrum, the outer surface atmosphere of the earth is no longer a barrier but a "window"; through this "window," the radiator on the ground can radiate directly into outer space. Usually, it is called "infrared window" in infrared technology.

The radiative thermal insulation coatings are a system with special fillers, which can convert the absorbed energy into molecular vibration and rotational energy; so the absorbed energy can be eventually transmitted to external space in the form of thermal radiation. Based on these, object covered with thermal radiation film can radiate more energy to external space than it absorbs from solar at the certain wavelength; as a result, the radiative thermal insulation film can cool the covered object actively. This thermal insulation mechanism in radiative coatings is quite different from the obstructive and reflective coatings mentioned above. As with either obstructive or reflective fillers, the film can only block extra solar energy passively, but with radiative fillers, the film can radiate the extra solar energy to external space actively.

As discussed above, radiative fillers showed excellent thermal radiation ability when the outer surface atmosphere of the earth is worked as a "window." So, in order to meet higher emissivity of film, fillers with strong absorption in the band from 8 to 13 μm are the key to coatings. Studies [35–39] have shown that adding a certain amount of far-infrared fillers into coating system can greatly enhance the infrared radiation ability of the film. Usually, Fe_2O_3, MnO_2, Cr_2O_3, TiO_2, SiO_2, Al_2O_3, La_2O_3, and CeO with high emissivity are usually used as thermal radiation functional fillers. Meanwhile, materials with antispinel structure doping from a variety of metal oxide doping can be used as thermal radiation functional fillers due to its higher energy emissivity, like ATO, ITO, etc. [40–42].

One word in all, radiative fillers are the key factor to achieve excellent thermal radiation in coatings; so the development on new radiative fillers in recent years promoted thermal insulation performance of coatings, but the thermal radiation ability of fillers is affected by many factors like the concentration, diameter size, surface properties (roughness/periodicity) of fillers, doping or not, and so on; so the main problem in radiative commercial coatings is that fillers with steady and excellent thermal insulation performance are expensive.

2.4 Composite thermal insulation coatings

As the heat transfer of object is a combination of heat conduction, convection, and radiation, the ideal thermal insulation coating can resist heat transfer, reflect, and radiate the solar energy actively. Although obstructive, reflective, or radiative thermal insulation coatings mentioned above have its own advantage in thermal insulation, the thermal insulation performance with just single mechanism cannot meet the desire for comprehensive thermal insulation; so under this background, composite thermal insulation coatings are designed to achieve a synergistic thermal insulation with obstructive, reflective, and radiative [43, 44]. For example, nanotitanium-oxide-modified hollow beads are used as functional filler; the film shows excellent thermal insulation performance due to the high reflectivity of the titanium oxide and low thermal conductivity of hollow beads at the same time [45]. A composite thermal insulation coating was prepared with obstructive, reflective, and radiative fillers together in Ref. [46]; testing results showed that the thermal insulation grade of the film is R-21.1, the TSR value of the film goes to 0.79, and the energy emissivity value is as high as 0.83. Data show that the film can resist heat transfer effectively, reflect most of the solar energy, and can cool the substrate actively by radiating energy absorbed. Multithermal insulation system with obstructive, reflective, and radiative fillers compatible with either acrylic or fluorocarbon substrate shows better thermal insulation performance than that with just single thermal insulation mechanism filler [47–49]. So composite thermal insulation coatings now became the main direction of thermal insulation coating research.

3. Application and development of thermal insulation coatings

Covering with thermal insulation coating has been one of the most effective techniques for energy saving. As discussed above, the thermal insulation performance of coatings is mainly affected by functional fillers, but the applicability of coatings is mainly affected by substrate. So when the coating is designed for a particular application, both thermal insulation performance and comprehensive performance like protective, decorative, and other special needs (anticorrosive, waterproof, fireproof, antifouling, conductive, sterilization, and so on) should be considered at the same time. In this situation, the multifunctional coatings with thermal insulation and other special functions can satisfy more to the requirements of market. Just take building energy saving for example, in order to achieve the overall thermal insulation effects, not only varies of thermal insulation coatings for the outside and inside walls of building have been produced. Meanwhile, considering the urgent thermal insulation needs on color steel plate, aluminum profiles, glass doors, and windows in modern building, researchers have been committed to the development of multifunctional coatings that can meet both the thermal insulation and other specific functional requirements of these structural components. To be specified, transparent thermal insulation coatings can be used for windows, thermal insulation, and anticorrosion coating coatings for aluminum profiles, and etc. Therefore, based on the practical application, the multifunctional coatings with thermal insulation and other special functions are the development trend for thermal insulation coatings.

3.1 Transparent thermal insulation coatings

Transparent thermal insulation coating is transparent in the visible light area with semiconductor powder as fillers. Materials with good transmittance on the visible spectrum and high infrared light transmittance can be used as functional fillers, including nanotin oxide antimony (ATO), nanoindium-tin oxide (ITO), etc., so the film with these fillers can show an excellent thermal insulation performance while being transparent [50–54].

Due to the unique size effect, localized field effect, quantum effect, and other unique properties, the nanoparticles can obviously improve both the thermal insulation and antiaging properties of the film. The transparent thermal insulation coatings can widely be used in glass doors and windows in modern buildings, automotive glass, and so on. In fact, transparent thermal insulation coatings can almost be used at any substrate with a particular need for both transmittance and thermal insulation needs.

A transparent thermal insulation coating with nano-ATO as filler was prepared and tested; results showed that the coatings show both good transparency and thermal insulation performance due to the use of nano-ATO. Moreover, the thermal insulation effect of the film increases with increasing weight content of ATO [52]. Test results also indicated that the transparent thermal insulation coatings with ATO possess good artificial accelerated weathering resistance.

3.2 Vacuum thermal insulation coatings

As thermal conduction caused by the molecular vibration and convection will completely disappear in vacuum, the thermal insulation performance of the film will be outstanding if the film can form a vacuum or near vacuum structure. In 1970s, experts in the United States obtained a high-quality thermal insulation coating, with aerogel as filler; the aerogel was prepared by filling spherical hollow ceramic microbubbles into an inert latex binder (aqueous) through NASA space-craft insulation material technology. The aerogel then forms a vacuum cavity layer in the film, which can not only obstruct but also reflect solar energy effectively. Tests showed that just brushing a thin layer of the film on the surface of buildings, the room temperature increased in winter but decreased in summer. That means the coating showed an effective thermal insulation effect [55]. Moreover, data show that the film can reach a thermal insulation up to 95%, and as a result, reduced up to 30–60% energy consumption when used on buildings. That means the vacuum thermal insulation coatings are excellent in both thermal insulation and compre-hensive performance due to its special structure [56, 57]. And it is considered to be one of the most efficient energy-saving materials with a promising future.

3.3 Nanoporous thermal insulation coatings

As mentioned above, aerogel with vacuum shows an ideal thermal insulation performance when used as fillers. But it is not easy to get a complete vacuum condi-tion in many situations. Under this situation, researchers tried to use aerogel alone as filler. Aerogel basically consists of ultrafine particles and gaseous dispersion medium. Usually, the particles are filled in the pores of the medium's network structure. It is found that when the pores in network are less than 50 nm, the aerogel can show a very good thermal insulation effect. Actually, the fillers' ideal thermal conductivity value can even approach zero. So, it is entirely possible to obtain a coating with smaller thermal conductivity value than that of static air (0.023 $W·m^{-1}·K^{-1}$) with fillers with nanoporous structure [58], which means a lot to thermal insulation performance of the film. So, fillers with nanoporous structure provided unprecedented opportunities and possibilities for the development of thermal insulation coatings.

Aerogels are low-density solid materials with nanoporous network structures. The aperture of SiO_2 aerogel is about 2–50 nm, and the hole rate is high up to 99.8%, and the thermal conductivity value of SiO_2 aerogel is 0.008–0.018 $W·m^{-1}·K^{-1}$ at room temperature, which is much lower than 0.023 $W·m^{-1}·K^{-1}$. So, SiO_2 aerogel is considered to be one of the lowest thermal conductivity materials in the field of thermal insulation. The thermal insulation performance of SiO_2 aerogel composites were also prepared and studied. For example, SiO_2 aerogel composed with ceramic fibers was studied. As discussed, silica aerogel itself has very low thermal conduc-tivity value on both gas and solid due to its special structure; meanwhile, ceramic fibers can greatly decrease the value of radioactive thermal conductivity of the composite. So as a result, silica aerogel composites show excellent thermal insula-tion properties. Test showed that the thermal conductivity value of the composite is only 0.017 and 0.042 $W·m^{-1}·K^{-1}$ accordingly when test at 200 and 800°C [59]. A trimethylchlorosilane-modified SiO_2 aerogel was prepared and tested, results indi-cated that the thermal conductivity of composite is 0.0136 and 0.0284 $W·m^{-1}·K^{-1}$ at room temperature and 400°C, respectively [60].

With SiO_2 aerogel and polyvinylidene fluoride as substrate, a thermal insulation film was prepared and tested. Results indicated that the thermal conductivity of the film is as low as 0.028 $W·m^{-1}·K^{-1}$ [61]. Meanwhile, the performance of thermal insulation enhanced with the increasing content of SiO_2 aerogel [62, 63].

3.4 Smart thermal insulation coatings

Smart thermal insulation coating, which can insulate heat when the outer temperature is too high and release heat when the outer temperature is too high, has drawn attention in recent years, as this kind of coating has both energy storage and thermal insulation functions. Thermochromic, photochromic, electrochromic, and gasochromic films are demonstrated for energy saving as different kinds of thermal insulation coatings [64–73]. By just taking thermo-chromic films for example, thermo-chromic materials are capable of changing their optical properties when exposed to heat. The transmittance and reflectance can be significantly altered due to phase transition. Metal oxides such as lower oxides of vanadium, titanium, iron, and niobium can be used as fillers, which means, with these fillers in coating system, the color of the film can change when the temperature changes. With lower transition temperature and sharp transition features, vanadium dioxide (VO_2)-based smart coatings have gained much attention in recent years. When the temperature is lower than 68°C (Tc), the structure of VO_2 is semiconducting (insulating) monoclinic phase; when the temperature exceeds Tc, the structure turns to metallic tetragonal rutile [74]. The switch between the different structures means different light selectivity, which means, at temperature under Tc, film with VO_2 allows transmission of the visible and infrared light, and when the temperature is higher than Tc, the VO_2 film allows visible light but blocks IR. As a result, film with VO_2 shows variation color when the temperature changes. Researchers have already done much to improve the luminescence transmittance and modulation capability of solar energy [75–78].

Based on the adjustability of the coating system, the study on smart thermal insulation coating causes more and more attention from the researchers; so it is worth looking forward to the widespread application of the smart thermal insulation coatings sometime in the future.

Acknowledgements

This chapter is supported by the key Project of Department of Education of Guangdong Province (2016GCZX008), the key Research Platform Project of Department of Education of Guangdong Province (gg041002) and the Project of Engineering Research Center of Foshan (20172010018).

Author details

Xiufang Ye* and Dongchu Chen
School of Materials Science and Energy Engineering, Foshan University, Foshan, Guangdong Province, China

*Address all correspondence to: yexf@fosu.edu.cn

IntechOpen

References

[1] Cai WG et al. China building energy consumption: Situation, challenges and corresponding measures. Energy Policy. 2009;**37**(6):2054-2059

[2] Al-Homoud DMS. Performance characteristics and practical applications of common building thermal insulation materials. Building and Environment. 2005;**40**(3):353-366

[3] Berardi U. A cross-country comparison of the building energy consumptions and their trends. Resources, Conservation and Recycling. 2017;**123**:230-241

[4] Goudarzi H, Mostafaeipour A. Energy saving evaluation of passive systems for residential buildings in hot and dry regions. Renewable and Sustainable Energy Reviews. 2017;**68**:432-446

[5] Berardi U. The outdoor microclimate benefits and energy saving resulting from green roofs retrofits. Energy and Buildings. 2016;**121**:217-229

[6] Powell MJ et al. Intelligent multifunctional $VO_2/SiO_2/TiO_2$ coatings for self-cleaning, energy-saving window panels. Chemistry of Materials. 2016;**28**(5):1369-1376

[7] Omrany H et al. Application of passive wall systems for improving the energy efficiency in buildings: A comprehensive review. Renewable and Sustainable Energy Reviews. 2016;**62**:1252-1269

[8] Bao Y et al. Monodisperse hollow TiO_2 spheres for thermal insulation materials: Template-free synthesis, characterization and properties. Ceramics International. 2017;**43**(12):8596-8602

[9] Ozel M. Thermal performance and optimum insulation thickness of building walls with different structure materials. Applied Thermal Engineering. 2011;**31**(17):3854-3863

[10] Cernuschi F et al. Modelling of thermal conductivity of porous materials: application to thick thermal barrier coatings. Journal of the European Ceramic Society. 2004;**24**(9):2657-2667

[11] Sun Z et al. Porous silica ceramics with closed-cell structure prepared by inactive hollow spheres for heat insulation. Journal of Alloys and Compounds. 2016;**662**:157-164

[12] Liao Y et al. Composite thin film of silica hollow spheres and waterborne polyurethane: Excellent thermal insulation and light transmission performances. Materials Chemistry and Physics. 2012;**133**(2):642-648

[13] Bouchair A. Steady state theoretical model of fired clay hollow bricks for enhanced external wall thermal insulation. Building and Environment. 2008;**43**(10):1603-1618

[14] Hu Y et al. Silicon rubber/hollow glass microsphere composites: Influence of broken hollow glass microsphere on mechanical and thermal insulation property. Composites Science and Technology. 2013;**79**:64-69

[15] Liang JZ, Li FH. Simulation of heat transfer in hollow-glass-bead-filled polypropylene composites by finite element method. Polymer Testing. 2007;**26**(3):419-424

[16] Chao Y et al. Surface modification of light hollow polymer microspheres and its application in external wall thermal insulation coatings. Pigment & Resin Technology. 2016;**45**(1):45-51

[17] Santamouris M, Synnefa A, Karlessi T. Using advanced cool materials in the

urban built environment to mitigate heat islands and improve thermal comfort conditions. Solar Energy. 2011;**85**(12):3085-3102

[18] Synnefa A, Santamouris M, Livada I. A study of the thermal performance of reflective coatings for the urban environment. Solar Energy. 2006;**80**(8):968-981

[19] Cheng V, Ng E, Givoni B. Effect of envelope colour and thermal mass on indoor temperatures in hot humid climate. Solar Energy. 2005;**78**(4):528-534

[20] Bansal NK, Garg SN, Kothari S. Effect of exterior surface colour on the thermal performance of buildings. Building and Environment. 1992;**27**(1):31-37

[21] Uemoto KL, Sato NMN, John VM. Estimating thermal performance of cool colored paints. Energy and Buildings. 2010;**42**(1):17-22

[22] Levinson R et al. Methods of creating solar-reflective nonwhite surfaces and their application to residential roofing materials. Solar Energy Materials and Solar Cells. 2007;**91**(4):304-314

[23] Levinson R, Akbari H, Reilly JC. Cooler tile-roofed buildings with near-infrared-reflective non-white coatings. Building and Environment. 2007;**42**(7):2591-2605

[24] Revel GM et al. Nanobased coatings with improved NIR reflecting properties for building envelope materials: Development and natural aging effect measurement. Cement and Concrete Composites. 2013;**36**:128-135

[25] Liu W et al. Facile synthesis and characterization of 2D kaolin/CoAl$_2$O$_4$: A novel inorganic pigment with high near-infrared reflectance for thermal

insulation. Applied Clay Science. 2018;**153**:239-245

[26] Qi Y, Xiang B, Zhang J. Effect of titanium dioxide (TiO$_2$) with different crystal forms and surface modifications on cooling property and surface wettability of cool roofing materials. Solar Energy Materials and Solar Cells. 2017;**172**:34-43

[27] Wang Z et al. A facial one-pot route synthesis and characterization of Y-stabilized Sb$_2$O$_3$ solar reflective thermal insulating coatings. Materials Chemistry and Physics. 2011;**130**(1):466-470

[28] YuXin CMJJC. Study of solar heat-reflective pigments in cool roof coatings. Journal of Beijing University of Chemical Technology (Natural Science Edition). 2009;**1**:013

[29] Shinkareva EV, Safonova AM. Conducting and heat-insulating paintwork materials based on nickel-plated glass spheres. Glass and Ceramics. 2006;**63**(1):32-33

[30] Wang MJ, Kusumoto N. Ice slurry based thermal storage in multifunctional buildings. Heat and Mass Transfer. 2001;**37**(6):597-604

[31] Stamatakis P. Optimum particle size of titanium dioxide and zinc oxide for attenuation of ultraviolet radiation. Journal of Coatings Technology. 1990;**62**(10):95

[32] Shen H, Tan H, Tzempelikos A. The effect of reflective coatings on building surface temperatures, indoor environment and energy consumption—An experimental study. Energy and Buildings. 2011;**43**(2):573-580

[33] Wang X et al. Dynamic thermal simulation of a retail shed with solar reflective coatings. Applied Thermal Engineering. 2008;**28**(8):1066-1073

[34] Guo W et al. Study on energy saving effect of heat-reflective insulation coating on envelopes in the hot summer and cold winter zone. Energy and Buildings. 2012;**50**:196-203

[35] He X et al. High emissivity coatings for high temperature application: Progress and prospect. Thin Solid Films. 2009;**517**(17):5120-5129

[36] Tan W, Petorak CA, Trice RW. Rare-earth modified zirconium diboride high emissivity coatings for hypersonic applications. Journal of the European Ceramic Society. 2014;**34**(1):1-11

[37] Neuer G, Jaroma-Weiland G. Spectral and total emissivity of high-temperature materials. International Journal of Thermophysics. 1998;**19**(3):917-929

[38] Huang J et al. Enhanced spectral emissivity of CeO$_2$ coating with cauliflower-like microstructure. Applied Surface Science. 2012;**259**:301-305

[39] Dan Z et al. Microstructure and properties of high emissivity coatings. Journal of University of Science and Technology Beijing, Mineral, Metallurgy, Material. 2008;**15**(5):627-632

[40] Granqvist CG, Hultåker A. Transparent and conducting ITO films: New developments and applications. Thin Solid Films. 2002;**411**(1):1-5

[41] Wang X et al. Effect of antimony doped tin oxide on behaviors of waterborne polyurethane acrylate nanocomposite coatings. Surface and Coatings Technology. 2010;**205**(7):1864-1869

[42] Cao XQ, Vassen R, Stoever D. Ceramic materials for thermal barrier coatings. Journal of the European Ceramic Society. 2004;**24**(1):1-10

[43] Cui W et al. Improving thermal conductivity while retaining high electrical resistivity of epoxy composites by incorporating silica-coated multi-walled carbon nanotubes. Carbon. 2011;**49**(2):495-500

[44] Xu L et al. Infrared-opacified Al2O$_3$–SiO$_2$ aerogel composites reinforced by SiC-coated mullite fibers for thermal insulations. Ceramics International. 2015;**41**(1, Part A): 437-442

[45] Gao Q et al. Coating mechanism and near-infrared reflectance property of hollow fly ash bead/TiO$_2$ composite pigment. Powder Technology. 2017;**305**:433-439

[46] Zhen Q et al. Development and application of thin layer coating of thermal insulation. Petroleum Engineering Construction. 2003;**29**(5):26-30

[47] Ye X et al. Preparation of a novel water-based acrylic multi-thermal insulation coating. Materials Science. 2017;**23**(2):173-179

[48] Ye XF et al. The preparation of fluorocarbon thermal insulation coating with different fillers. Advanced Materials Research. 2015;**1101**:36-39

[49] Zhang W et al. A systematic laboratory study on an anticorrosive cool coating of oil storage tanks for evaporation loss control and energy conservation. Energy. 2013;**58**:617-627

[50] Goebbert C et al. Wet chemical deposition of ATO and ITO coatings using crystalline nanoparticles redispersable in solutions. Thin Solid Films. 1999;**351**(1):79-84

[51] Guzman G et al. Transparent conducting sol–gel ATO coatings for display applications by an improved dip coating technique. Thin Solid Films. 2006;**502**(1):281-285

[52] Qu J et al. Transparent thermal insulation coatings for energy efficient glass windows and curtain walls. Energy and Buildings. 2014;**77**:1-10

[53] Yao L et al. Hard and transparent hybrid polyurethane coatings using in situ incorporation of calcium carbonate nanoparticles. Materials Chemistry and Physics. 2011;**129**(1-2):523-528

[54] Ghosh SS, Biswas PK, Neogi S. Thermal performance of solar cooker with special cover glass of low-e antimony doped indium oxide (IAO) coating. Applied Thermal Engineering. 2017;**113**:103-111

[55] Ruben B et al. Aerogel insulation for building applications. 2011;**43**(4):761-769

[56] He Y-L, Xie T. Advances of thermal conductivity models of nanoscale silica aerogel insulation material. Applied Thermal Engineering. 2015;**81**:28-50

[57] Tang G et al. Thermal transport in nano-porous insulation of aerogel: Factors, models and outlook. Energy. 2015;**90**:701-721

[58] Wu HJ, Fan JT, Du N. Porous materials with thin interlayers for optimal thermal insulation. International Journal of Nonlinear Sciences and Numerical Simulation. 2009;**10**(3):291

[59] Feng J et al. Preparation and properties of fiber reinforced SiO_2 aerogel insulation composites. Journal of National University of Defense Technology. 2010;**1**:009

[60] Kwon Y-G et al. Ambient-dried silica aerogel doped with TiO_2 powder for thermal insulation. Journal of Materials Science. 2000;**35**(24):6075-6079

[61] Wu H et al. Synthesis of flexible aerogel composites reinforced with electrospun nanofibers and

microparticles for thermal insulation. Journal of Nanomaterials. 2013;**2013**:10

[62] Reim M et al. Silica aerogel granulate material for thermal insulation and daylighting. Solar Energy. 2005;**79**(2):131-139

[63] Liu CL et al. Preparation of thin-film nano-scale thermal insulation coatings for exterior wall. Coatings Technology & Abstracts. 2014;**35**(7):15-18

[64] Zhu J et al. Vanadium dioxide nanoparticle-based thermochromic smart coating: High luminous transmittance, excellent solar regulation efficiency, and near room temperature phase transition. ACS Applied Materials & Interfaces. 2015;**7**(50):27796-27803

[65] Pause B. Development of heat and cold insulating membrane structures with phase change material. Journal of Coated Fabrics. 1995;**25**(1):59-68

[66] Loquai S et al. HiPIMS-deposited thermochromic VO_2 films with high environmental stability. Solar Energy Materials and Solar Cells. 2017;**160**:217-224

[67] Li M et al. Active and dynamic infrared switching of VO_2 (M) nanoparticle film on ITO glass. Journal of Materials Chemistry C. 2016;**4**(8):1579-1583

[68] Rezaei SD, Shannigrahi S, Ramakrishna S. A review of conventional, advanced, and smart glazing technologies and materials for improving indoor environment. Solar Energy Materials and Solar Cells. 2017;**159**:26-51

[69] Wu C-C, Shih W-C. Development of a highly transparent, low-resistance lithium-doped nickel oxide triple-layer film deposited by magnetron sputtering. Chemical Communications. 2017;**53**(10):1634-1637

[70] Lin T-C, Huang W-C, Tsai F-C. Hydrogen plasma effect toward the AZO/CuCr/AZO transparent conductive electrode. Microelectronic Engineering. 2017;**167**:85-89

[71] Hong-Tao S et al. Optimization of TiO$_2$/Cu/TiO$_2$ multilayers as a transparent composite electrode deposited by electron-beam evaporation at room temperature. Chinese Physics B. 2015;**24**(4):047701

[72] Kim JH et al. Dependence of optical and electrical properties on Ag thickness in TiO$_2$/Ag/TiO$_2$ multilayer films for photovoltaic devices. Ceramics International. 2015;**41**(6):8059-8063

[73] Seyfouri MM, Binions R. Sol-gel approaches to thermochromic vanadium dioxide coating for smart glazing application. Solar Energy Materials and Solar Cells. 2017;**159**:52-65

[74] Wang Y, Runnerstrom EL, Milliron DJ. Switchable materials for smart windows. Annual Review of Chemical and Biomolecular Engineering. 2016;**7**:283-304

[75] Chang T et al. Facile and low-temperature fabrication of thermochromic Cr$_2$O$_3$/VO$_2$ smart coatings: Enhanced solar modulation ability, high luminous transmittance and UV-shielding function. ACS Applied Materials & Interfaces. 2017;**9**(31):26029-26037

[76] Zhou L et al. Enhanced luminous transmittance of thermochromic VO$_2$ thin film patterned by SiO$_2$ nanospheres. Applied Physics Letters. 2017;**110**(19):193901

[77] Wang N et al. One-step hydrothermal synthesis of rare earth/W-codoped VO$_2$ nanoparticles: Reduced phase transition temperature and improved thermochromic properties. Journal of Alloys and Compounds. 2017;**711**:222-228

[78] Long S et al. Thermochromic multilayer films of WO$_3$/VO$_2$/WO$_3$ sandwich structure with enhanced luminous transmittance and durability. RSC Advances. 2016;**6**(108):106435-106442

Selection of the Best Optimal Operational Parameters to Reduce the Fuel Consumption Based on the Clustering Method of Artificial Neural Networks

Tien Anh Tran

Abstract

The international shipping transportation industry becomes gradually important in the field of national economic development. It is explained by means of an
increase in a number of ships nowadays, and it also expands the operating routes
on international routes including the North of America, Baltic Sea, and emission
control areas (ECAs). The energy efficiency of ships is very necessary to respond
to the regulations of the International Maritime Organization (IMO). Moreover,
the operational parameters have a significant meaning in supervising and monitoring the engines on a ship. They completely depend on the navigation environment
condition. So, selecting the optimal operational parameters' component is a target
of this study. In this chapter, a study on the energy efficiency of ship by decreasing the fuel consumption of the main engine for a certain ship namely M/V NSU
JUSTICE 250,000 DWT of VINIC shipping transportation company in Vietnam
is by the method of artificial neural networks (ANNs). In particular, these studies
were conducted by the classification and clustering method of artificial neural
networks (ANNs) based on the experimental database of M/V NSU JUSTICE
250,000 DWT. The results of this chapter will solve the energy efficiency issue on
ships nowadays and contribute the aims in the next studies.

Keywords: energy efficiency, fuel consumption of engines, artificial neural
networks, clustering, classification

1. Introduction

The international shipping transportation industry plays an important role in the
economic development of each nation. The benefits have been brought dramatically
through years. It is adopted that the number of cargoes that have been transported
between the ports increases; however, the development of shipping transportation
leads to the environmental pollution. This is a reason that the International Maritime
Organization (IMO) is adopted and conducted to solve the problems that concern
about the ship operation in the field of development of the international shipping
transportation industry.

The definition of a sustainable shipping transportation is identified through some fields such as maritime safety, sea environmental protection, ship energy efficiency management, security and ocean resource conservation, so on. Moreover, the ship energy efficiency management is a main part of issue in the field by decreasing the CO_2 gas emissions from the international shipping transportation industry. Hence, it is a big factor that contributes the climate changes and environmental pollution nowadays.

Following the data statistic, the harmful gases emit into the environment about 900 million tons of carbon dioxide in 2018. Amount of this gas gradually increases comparing each year [1]. Moreover, the CO_2 gas emission is the main cause to make the greenhouse gas (GHG) emission. The global warming phenomenon and climate changes are the serious effects that the world is facing today (**Figure 1**).

The International Maritime Organization (IMO) has made some progress, and the current debate is addressing how much the sector can be expected to reduce emissions and should be obliged to reduce, as well as in what manner these diminutions can be achieved [2]. On the other hand, the energy efficiency issue plays a vital role in decreasing the fuel oil consumption of main engine and equipment along with the restricting greenhouse gas emission, especially the carbon dioxide gas. This is leading to the energy efficiency issue, which becomes a mandatory for countries obeying the IMO's regulations and rules about the environmental protection. Especially, the main cause that makes this phenomenon is generated from the combustion chamber of engines on board. Marine diesel engines are self-ignition engines in heavy-duty vehicles, but they are generally larger in size, equipped with more complex system and operated with higher efficiency.

As an alternative, a simulation model can be developed to predict the actual condition of engine performance through the fuel consumption level of engines and navigation environmental condition impacts. Moreover, there are also a lot of recent researches about energy efficiency of ships by reducing the fuel consumption of main engine and equipment by applying the modern control theory, machine learning, or artificial neural networks. The artificial neural network method is used as an alternative method comparing with other traditional methods [3]. On the other hand, the input signal will be trained when admitting the artificial neural network. After that, the appropriate data and method will be used to obtain the best prediction. Finally, the output signal will be given out into result part.

On another side, the artificial neural network (ANN) method has been developed during many years with the aim of dealing the complex issues. Hence, the model is applied by ANNs and is possible to deal with other analytical and statistical methods [4]. Especially, the capability of forecast is approached by ANNs [3]. Besides that, the ANNs could fit the great adaptability, robustness, and major fault tolerance in case of highly processing factors [5]. Moreover, the surface fitting of ANNs will be applied in the study and favor the method in the field of establishing the prediction model [6].

There are a large number of studies that are applied in the field of designing the fuel oil consumption model of diesel engine by the ANN method [7–10]. ANN has been found to be the domain for many successful applications of prediction tasks, in modeling and prediction of energy-engineering system [11], prediction of the energy consumption of passive solar buildings [12], and analysis of the reduction in emissions [13]. In this study, the author has investigated the artificial neural network in particular the clustering data method in the field by reducing fuel consumption of main engine for bulk carriers when considering the navigation environment condition impacts. On the other hand, the data clustering method is carried out using big data from experimental data, then the research results are compared with the actual experimental data with the aim of regulating the

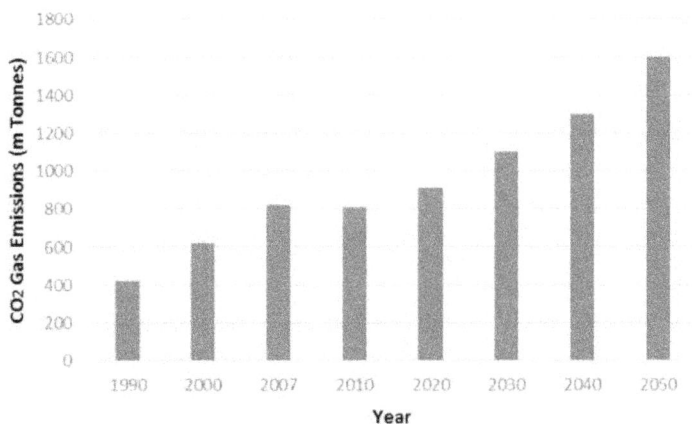

Figure 1.
The carbon dioxide (CO_2) emission the following years [1].

proper operation of ship and gaining the low fuel consumption of main engine on bulk carriers. Moreover, the study object is applied and verified throughout the certain bulk carrier, namely, M/V NSU JUSTICE 250,000 DWT of VINIC Shipping Transportation Company in Vietnam.

2. Literature review

2.1 Fuel oil consumption of diesel engine

Each kind of diesel engine has the specific fuel oil consumption. The fuel consumption of diesel engine will decide the working characteristics of each diesel engine and be able to generate the power and performance to the screw propeller. Almost diesel engines are equipped on ships that are two-stroke low-speed, large-size diesel engine of MAN B&W, Sulzer manufacturers. They are served as marine main diesel engine on ships and set-up on bulk carriers. Recently, there are some researches of scientist and researchers which investigate the fuel oil consumption of main diesel engine. Tran has proposed fuel oil consumption model of diesel engine when sailing on emission control areas (ECAs) by the artificial neural network [14]. The evaluation of ship engine effective power fuel consumption as well as gas emissions has been carried out by Borkowski et al. through ship's speed [15].

2.2 EEOI measure

EEOI—energy efficiency operational indicator—is a main parameter in operational energy efficiency measure of ships. The regulations of EEOI are defined in Chapter IV, Annex VI, MARPOL 73/78. Furthermore, there are some recent researches which concentrate on this index through their studies. A tool of EEOI calculation for bulk carriers of VINIC shipping transportation is carried out by Tran [16]. Hence, the optimization of this index is also conducted by him [17]. Consequently, the energy efficiency management of ships plays an important role in key policy strategy nowadays. The study of barriers in the field of ship's energy efficiency management has been conducted by Rehmatulla and Smith [18]. The study of the uncertainty hull form optimization method has been investigated by Hou with the aim of lowering the EEOI index [19].

2.3 Data analysis methods

2.3.1 Artificial neural networks (ANNs)

In recent times, the artificial neural networks (ANNs) have been studied and applied in many fields of science and technology. Especially, it has been used in the field of the data clustering method through analyzing the architecture and pattern recognition. In case of pattern recognition, this method has presented the input and output nodes in which they are linked each other with differential weights. The proposed model will have a mission, which creates the relationship between input node and output node. This relationship will be adjusted until a termination criterion is satisfied. This process of weight adjustment, called learning, lends continuous learning or artificial learning capability to the system, which can be either supervised or unsupervised learning capability to the system, which can be either supervised or unsupervised learning in ANN. The supervised learning demands an output class declaration for each of the inputs. The unsupervised learning network itself recognizes the features of the input and self organizes the inputs. The parametric and nonparametric approach will be reached. The parametric approach will be combined between classification and parameterization. The nonparametric approach will include the unclassified data that used the adaptive resonance theory (ART) method. This combination will be based on neurophysiology including prior knowledge and adaptive resonance theory (ART) method. Hence, this one will be known as stability plasticity dilemma.

The basic block of artificial neural network model is an artificial neuron. Each artificial neuron has three sets of rule including multiplication, summation, and activation. Each input node of artificial neural network will have multiple weight values. The weight value has gained the separate function of artificial neural network. The definition of transfer function is known as the weighted sum of previous input nodes and bias (**Figure 2**) [20]. A simple neural network is known as a real power when it is connected with other neurons in the same network. The progress of dealing will be reviewed all neurons through transferring information between nodes each other. The equation of transferring information will be represented as [20]:

$$y(k) = F\left(\sum_{i=0}^{m} w_i(k).x_i(k) + b\right) \tag{1}$$

where $x_i(k)$ is the input value in discrete time k, and i ranges from 0 to m; $w_i(k)$ is the weight value in discrete time k, and i ranges from 0 to m; b is the bias; F is the transfer function; and y (k) is the output value in discrete time k.

Competitive learning exists in biological neural networks. Competitive or winner-take-all neural networks [21] are used often to cluster input data. The characteristic of the same pattern is grouped, which is represented by a single unit. This group will be hanged automatically on the same basis data. The process of weight update will be carried out and divided into a certain group. The Kohonen's Learning Vector Quantization (LVQ) and Self-Organizing Map (SOM) use for familiar artificial neural networks [22] and adaptive resonance theory models [23]. The two-dimensional map of multidimensional data has been used for vector quantization and speech recognition [22]. In addition, the learning rate and a neighborhood of win node have been studied and controlled. Carpenter and Grossberg [23] use ART model in order to support more stably and more plastic. The partitions are approached for different ones. Moreover, the ART net will be made up of the number and size of clusters. The pattern will be classified into different groups by vigilance threshold. Hence, the hyper spherical cluster is fit for both SOM and ART [24].

2.3.2 Data clustering method

The data analysis technique has been investigated in this research. This technique is useful in the field of big data. The research key of the data clustering method is a classification of big data into a separate group in which each group will contain the data that is the same characteristic together. Even though there is an increasing interest in the use of clustering methods in pattern recognition [16], image processing [17], and information retrieval [18, 19], clustering has a rich history in other disciplines [20] such as biology, geography, geology, archeology, psychology, psychiatry, marketing, and finance.

The data clustering method is analyzed through input data. Krenker and his colleague [20] had the debate section between the data clustering methods that includes pattern recognition, classification, and clustering data. The study of the fuzzy set theory technique has been carried out in the process of classification and robust approach. The machine learning technique includes the artificial neural network (ANN), the nonlinear characteristics of data, and the classification of data.

2.3.3 Data analysis through ANNs

The data analysis is a complex subject in the field of machine learning. This field is used in the complex structure models. Normally, the data usually have a certain rule. So, the clustering method is used as an effective method in the field of controlling data. This tool includes an unsupervised classification technique. It presents some inherent structures in data set. All the classification steps will look up from an appropriate function with data groups in the proposed method [25, 26]. In addition, a neural network is a nonlinear control model that is based on the real complex process. This model provides the basic classification rule and statistical data analysis [27]. The neural network has been used as a potential alternative method in the field of the classification method. This method has been confirmed as a useful technique for data clustering. So, the output layers have been considered as a competition layer. The competition layer will be connected together along with the input layers [22, 23].

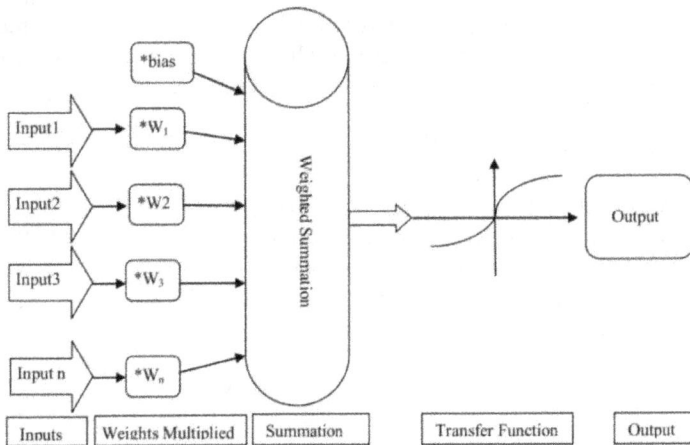

Figure 2.
Artificial neural network simple model.

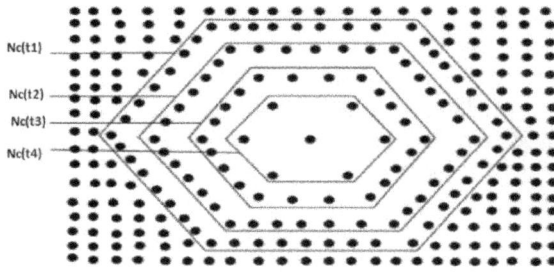

Figure 3.
Topological neighborhood of Kohonen's net Nc(t).

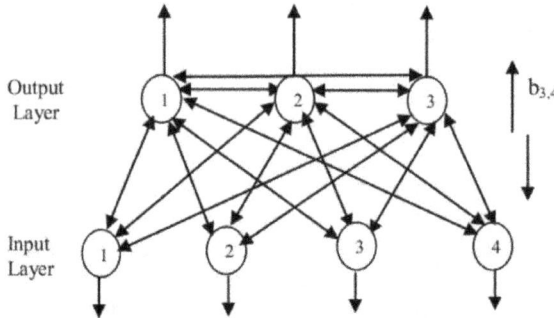

Figure 4.
ART1 networks.

2.3.4 Self-organizing feature maps (SOFMs)

Self-organizing feature maps (SOFMs) also called Kohonen feature maps [22] are laid on the category of learning clustering. The nodes of artificial neural network become various input nodes. It concludes two layers of neuron: input layer and competition layer. The winner neuron is determined into competition layer. The competition process is based on the clustering method not only the weight of winner neuron but also its neighborhood Nc(t) (with t1 < t2 < t3 < t4), which is defined in terms of some proximity relation. This neighborhood relation is usually represented as a grid (usually two dimensional) in **Figure 3**.

2.3.5 Adaptive resonance theory (ART)

Adaptive resonance theory is presented as the ART method in which it stands for the input vector with an active code vector. The first ART model, ART1, given by Carpenter and Grossberg [28] is shown in **Figure 4**.

ART networks are based on Stephen Grossberg's stability plasticity dilemma and are supposed to be stable and plastic [23].

2.3.6 Learning vector quantization (LVQ)

Learning vector quantization (LVQ) is known as an effective method for training the competition layers of neural network artificial. An LVQ has also the same architecture which is expected into classical membership function. Each LVQ network will have competition layer and linear layer. Both competition layer and linear layer will have one neuron per class. In case of two categories of LVQ models, the supervised mode will consist of LVQ1, LVQ2, and LVQ3 [22] and unsupervised like LVQ4 and incremental c means [29].

3. The case study

3.1 M/V NSU JUSTICE 250,000 DWT

The target ship is chosen in this study, a kind of bulk carrier with certain name M/V NSU JUSTICE 250,000 DWT. This is the largest ship of VINIC shipping transportation company, Haiphong, Vietnam. The main routes of this vessel concentrate on the international routes such as Japan-Australia-Brazil-the United States-emission control areas (ECAs; **Figures 5** and **6**).

In **Table 1**, the technical parameters of the main propulsion plant on M/V NSU JUSTICE 250,000 DWT are shown. They definitely concern about the fuel consumption level of the main engine and operational working condition of ship.

3.2 Data clustering method for bulk carrier

From the above theoretical research, the data clustering method based on artificial neural networks (ANNs) plays a vital role in data mining, and it is applied effectively for the ships with large size and complex routes. In this research, the author carried out researching the clustering method with applied object is a certain vessel of VINIC shipping transportation company in Vietnam. M/V NSU JUSTICE 250,000 DWT is the largest size of company equipped with nine cargo holds. This is a kind of bulk carrier, and it usually operates on Japan, Australia, and Brazil. The experimental data were collected from completed voyages at certain different ports in 1 month. **Table 2** shows the separate voyages of M/V NSU JUSTICE 250,000 DWT. The order item is corresponding to the in turn of the voyage number 16, 17, 18, 19, and 20.

In **Table 2**, the input parameters including the average draft of M/V NSU JUSTICE have conducted the voyage from the departure port to the destination one. Each port has different drafts depending on the density of water and water temperature and weather condition at certain time. The working hours and propeller hours under water also referred as the input parameters. Voyage distance is also listed. On the other hand, the speed of the ship, main shaft revolution, and ship speed slip degree are also included in the input parameter component. Especially, the fuel

Figure 5.
M/V NSU JUSTICE 250,000 DWT.

Figure 6.
Main diesel engine (MAN B&W 7S80MC-C).

Ship name	NSU JUSTICE	Type of main engine	MAN B&W 7S80MC-C
IMO N°	9,441,922	Maximum continue rating (MCR)	21,910 kW
Flag	PANAMA	Revolution per minute (100% MCR)	74.5 rpm
Built year	2012	Fuel oil consumption	160.9 g/kWh
Dead weight tons (DWT)	250,000	Number of cylinders	7
Length (m)	329.95	Cylinder bore (mm)	800
Width (m)	57.00	Piston stroke (mm)	3200
Draft (m)	18.00	Turbocharger	Axial flow

Table 1.
The specification of M/V NSU JUSTICE and main diesel engine.

No.	Voyage no.	Draft	Working hours (Hrs)	Propeller hours (Hrs)	Distance (Knots)	Ship speed (Knots/h)	Speed slip	Shaft revolution (rpm)	FOC (MT)
1	16	9.95	270.25	264.5	3458	13.1	−0.2	59.7	512.2
2	17	9.67	271.75	268.5	4053	15.3	−1.2	69.8	747.7
3	18	9.82	826.5	809.5	12,194	15.1	0.1	69.7	2256.6
4	19	9.71	234.75	230	3463	15	0.2	69.7	645.7
5	20	9.68	233.18	229.8	3431	15	−0.8	69.1	629.3
FOC = fuel oil consumption.									

Table 2.
Operational data of M/V NSU JUSTICE 250,000 DWT.

consumption of main engine is the target input that needs to reduce with the aim of solving the energy efficiency of ships. In this research, the energy efficiency of ship proposes to the bulk carriers, particularly M/V NSU JUSTICE 250,000 DWT of VINIC shipping transportation company in Vietnam.

4. Results and discussion

Throughout the neural network toolbox on MATLAB program, the working of data clustering was completely conducted based on the artificial neural networks (ANNs). The algorithm of data clustering is self-organizing map (SOM), and it is explained that the same characters will be classified at the same cluster. The benefit of data clustering method will give out the proper values in series of the experimental data of this vessel, M/V NSU JUSTICE 250,000 DWT.

The self-organizing map topology in the clustering method based on the artificial neural network toolbox is shown in **Figure 7**.

In **Figure 8**, the SOM neighbor connections are indicated based on the self-organizing map data clustering. In this figure, each node has the connection together. Following this, the SOM neighbor connection that will be represented through each node will have a certain connection. This connection has been explained clearly according to the data clustering method of artificial neural network theory. The material data will be added into this model and analyzed the following certain rules.

On another side, the distance of weights in self-organizing map (SOM) data clustering is also shown in **Figure 9**. It helps the users to recognize the weight position in grid map along with the distance of different weights.

Each weight is distributed in different positions corresponding to **Figures 10–14**. Based on the proposed method, the operational data of M/V NSU JUSTICE 250,000 DWT have been analyzed and clustered into a certain group. From the operational data of M/V NSU JUSTICE 250,000 DWT, these data have been divided into a certain group through the data clustering method of artificial neural network in MATLAB program. In particular, the operational data have been divided into five inputs through **Figures 10–14**.

In a result, the best optimal operational parameters for the energy efficiency of M/V NSU JUSTICE 250,000 DWT laid on weights 1 and 2 corresponding to the minimum of the main engine fuel consumption level. It means that the operators need to remain the ship's speed at a certain level, and then the fuel consumption is lower when changing the main engine working conditions. The optimal value has been selected according to the initial condition and certain rule of data clustering method of artificial neural networks. There are five voyages, and then, there are two voyages, which are selected as a specific example of the data clustering method in the field of data analysis.

In **Table 3**, the best parameters on M/V NSU JUSTICE 250,000 DWT are shown.

It is suggested that for the bulk carriers in general should remain the ship's speed in the range of 13–15 knots/h then assuring the navigation elements (i.e., just in time, a weight of cargoes carried, etc.) and the energy efficiency of ships (**Figure 15**).

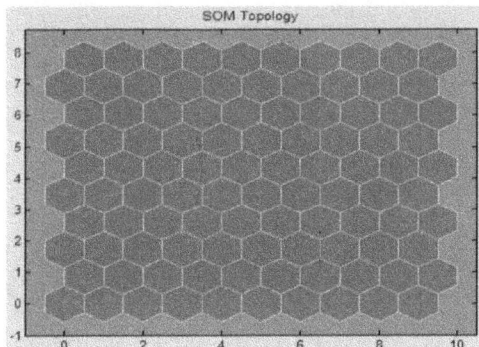

Figure 7.
SOM topology for experimental data.

In order to identify this effective proposed method, the author has conducted to compare the data clustering method of artificial neural network with the fuzzy clustering method same operational data of M/V NSU JUSTICE 250,000 DWT. The results of data clustering are shown in **Figure 16** [30].

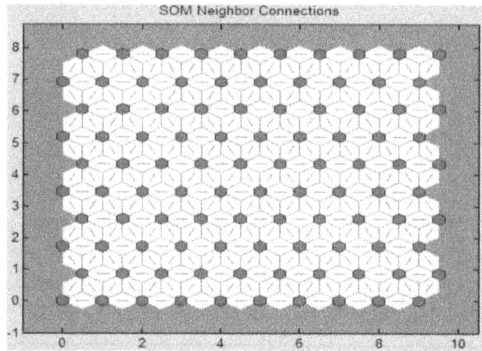

Figure 8.
SOM neighbor connections.

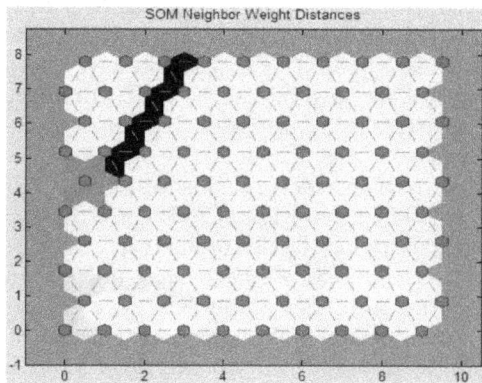

Figure 9.
SOM neighbor weight distances.

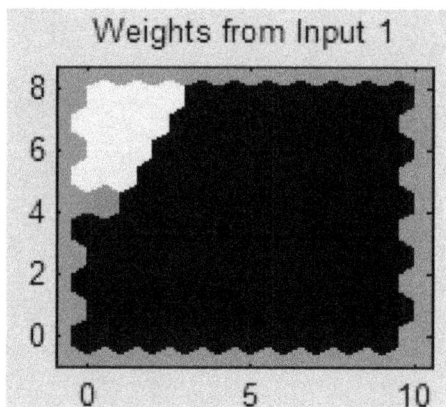

Figure 10.
Weight density of Input 1.

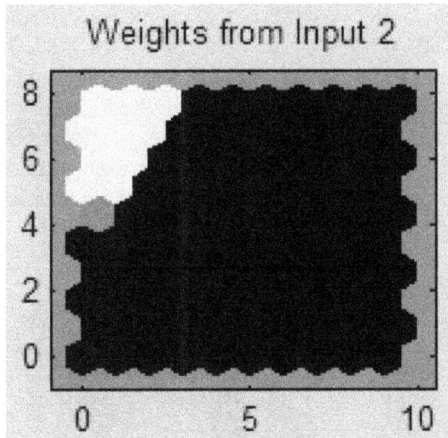

Figure 11.
Weight density of Input 2.

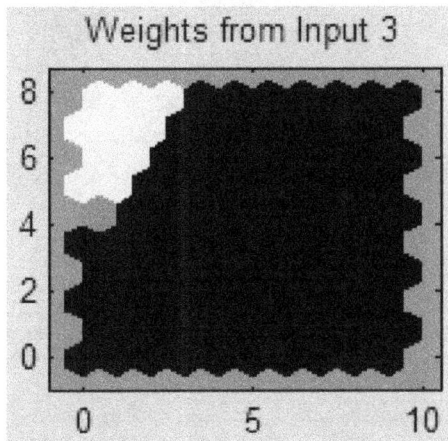

Figure 12.
Weight density of Input 3.

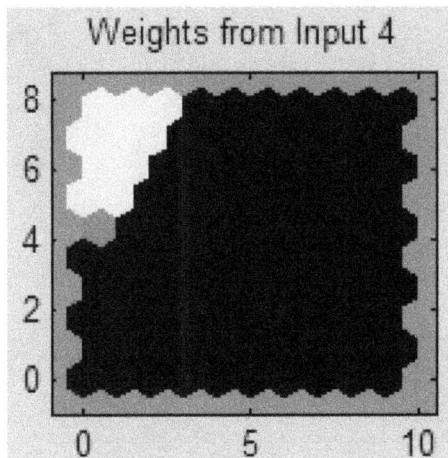

Figure 13.
Weight density of Input 4.

In reality, the data clustering is carried out by the fuzzy clustering method. The operational data of M/V NSU JUSTICE 250,000 DWT have been divided into a certain group. In this case, there are three groups. However, this method is limited with parameters of input values. There are four parameters including wind speed, wave height, fuel oil (FO) consumption, and diesel oil (DO) consumption. And then, the data fuzzy clustering method of artificial neural network can deal with various parameters of navigation environment conditions and selected the optimal voyage of vessel corresponding to the lowest fuel oil consumption of main diesel engine of bulk carriers.

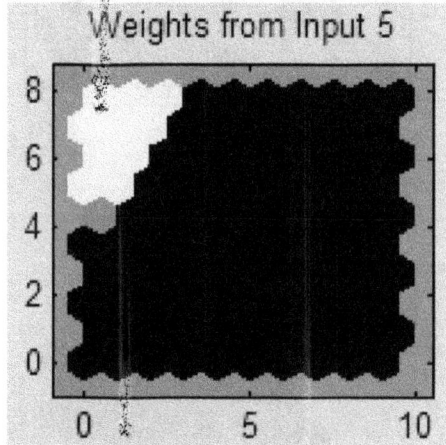

Figure 14.
Weight density of Input 5.

No.	Voyage no.	Draft	Working hours (Hrs)	Propeller hours (Hrs)	Distance (Knots)	Ship speed (Knots/h)	Speed slip	Shaft revolution (rpm)	FOC (MT)
1	16	9.95	270.25	264.5	3458	13.1	−0.2	59.7	512.2
2	17	9.67	271.75	268.5	4053	15.3	−1.2	69.8	747.7

Table 3.
The best optimal operational parameters.

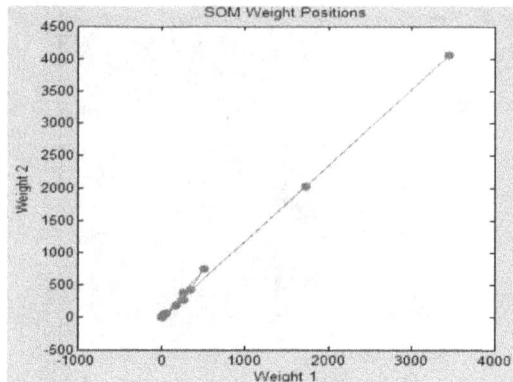

Figure 15.
SOM weight positions.

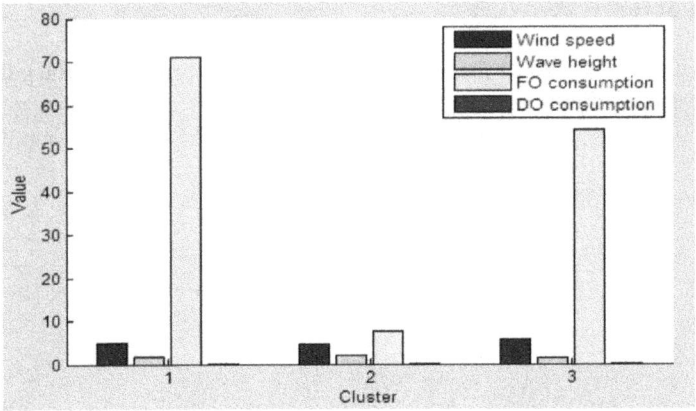

Figure 16.
Data clustering based on the fuzzy clustering method for M/V NSU JUSTICE 250,000 DWT [30].

5. Conclusions

The energy efficiency of ships has important properties in creating the green shipping nowadays. It is not only trending all ships following the International Maritime Organization (IMO)'s regulations, especially the International Convention for the Prevention of Pollution from Ships (MARPOL 73/78), Annex VI, Chapter 4 but also rising the economical effectivities for ship owners and ship operators. The operational parameters are important elements in order to decide the energy efficiency of ships. Hence, the selection of best optimal operational parameters plays an important role in the field by reducing the fuel consumption but ensuring the working ability of propulsion plants on ship, especially the main engine. Through the results of this research, the use of the clustering method in ANN has dealt with data analysis issue when ship operators are facing with the numerous operational data collected from voyages. Throughout the research results, the use of data clustering of the artificial neural network method can be selected by the optimal parameters in order to save the fuel oil consumption of main diesel engine for bulk carriers of VINIC shipping transportation company in Vietnam. These parameters will decide directly to fuel oil consumption of vessel. The navigation environment conditions, working hours, sailed distance, ship speed, and propeller speed split impact on fuel oil consumption of main diesel engine. The comparison between proposed method with other traditional methods then the data clustering method of artificial neural network has been presented more clearly through this research. This method has restricted some disadvantages of traditional data analysis methods. The data clustering quality clearly increases and determines the optimal voyage in the field by decreasing the fuel oil consumption of main diesel engine. Moreover, this research will bring effectively in saving the fuel consumption of main diesel engine and improving the ship's energy efficiency management in the shipping transportation industry.

Acknowledgements

The author would like to thank Dr. Murat Eyvaz, Dr. Ebubekir Yuksel, and Dr. Abdulkerim Gok for commencing and reviewing this chapter of our book. Additionally, the author acknowledges chief engineer Hoang Van Thuy of M/V NSU JUSTICE for his suggestion.

Conflict of interest

The author declares that there is no conflict of interest regarding the publication of this research chapter.

Author details

Tien Anh Tran
Faculty of Marine Engineering, Vietnam Maritime University, Haiphong, Vietnam

*Address all correspondence to: trantienanhvimaru@gmail.com

IntechOpen

References

[1] IMO. Second IMO GHG Study. London: International Maritime Organization; 2009

[2] Faber J et al. Schwarz Technical Support for European Action to Reducing Greenhouse Gas Emissions from International Transport. The Netherlands: Oude Delft 1802611 HH Delft; 2009

[3] Zhang G et al. Forecasting with artificial neural networks: The state of the arts. International Journal of Forecasting. 1998;**4**(1):35

[4] Hornik K et al. Multilayer Feed Forward Networks are Universal Approximators. Oxford, UK: Elsevier Science Ltd.; 1989. p. 5

[5] Lippman RP. An introduction to computing with neural nets. IEEE ASSP Magazine. 1987;**4**(2):4-22

[6] He X et al. Automatic sequence of 3D point data for surface fitting using neural networks. Computers and Industrial Engineering. 2009;**57**(1):408-418

[7] Togin NK, Baysec S. Prediction of torque and specific fuel consumption of a gasoline engine by using artificial neural networks. Applied Energy. 2010;**87**:349-355

[8] Ajdadi FR, Gilandeh YA. Artificial neural netwok and stepwise multiple range regression methods for prediction of tractor fuel consumption. Measurement. 2011;**44**:2104-2111

[9] Wu JD, Liu JC. Development of a predictive system for car fuel consumption using an artificial neural network. Expert Systems with Applications. 2011;**38**:9467-9471

[10] Safa M, Samarasinghe S. Modeling fuel consumption in wheat production using artificial neural networks. Energy. 2013;**49**:337-343

[11] Kalogirou SA. Applications of artificial neural networks for energy systems. Applied Energy. 2000;**67**:17-35

[12] Kalogirou SA, Bojic M. Artificial neural networks for the prediction of the energy consumption of a passive solar-building. Energy. 2000;**25**:479-491

[13] Amirnekooei K et al. Integrated resource planning for Iran: development of reference energy system, forecast, and long-term energy-environment plan. Energy. 2012;**46**:374-385

[14] Tran TA. Design the prediction model of low-sulfur-content fuel oil consumption for M/V NORD VENUS 80,000 DWT sailing on emission control areas by artificial neural networks. Proceedings of the Institution of Mechanical Engineers, Part M. 2017:1-18. DOI: 10.1177/1475090217747159

[15] Borkowski T et al. Assessment of ship's engine effective power fuel consumption and emission using the vessel speed. Journal of KONES. Powertrain and Transport. 2011;**18**(2):31-39

[16] Tran TA. A research on the energy efficiency operational indicator EEOI calculation tool on M/V NSU JUSTICE of VINIC transportation company, Vietnam. Journal of Ocean Engineering and Science. 2017;**2**(1):55-60

[17] Tran TA. Optimization of the energy efficiency operational indicator for M/V NSU JUSTICE 250,000 DWT by grey relational analysis method in Vietnam. Proceedings of the Institution of Mechanical Engineers, Part M. 2017:1-15. DOI: 10.1177/1475090217748756

[18] Rehmatulla N, Smith T. Barriers to energy efficiency in shipping: A triangulated approach to investigate the principal agent problem. Energy Policy. 2015;**84**:44-57

[19] Hou YH. Hull form uncertainty optimization design for minimum EEOI with influence of different speed perturbation types. Ocean Engineering. 2017;**140**:66-72

[20] Krenker A et al. Bidirectional artificial neural networks for mobile-phone fraud detection. ETRI Journal. 2009;**31**(1):92-94

[21] Jain AK, Mao J. Artificial neural networks: A tutorial. IEEE Computer. 1996;**29**(Mar.):31-44

[22] Kohonen T. Self-Organization and Associative Memory. In: Springer Information Sciences Series. 3rd ed. New York, NY: Springer-Verlag; 1989

[23] Carpenter G, Grossberg S. ART3: Hierarchical search using chemical transmitters in self-organizing pattern recognition architectures. Neural Networks. 1990;**3**:129-152

[24] Hertz J et al. Introduction to the Theory of Neural Computation. Santa Fe Institute Studies in the Sciences of Complexity Lecture Notes. MA: Addison-Wesley Longman Publ. Co., Inc., Reading; 1991

[25] Cybenko G. Approximation by superpositions of a sigmoidal function. Mathematics of Control, Signals, and Systems. 1989;**2**:303-314

[26] Hornik K. Approximation capabilities of multilayer feedforward networks. Neural Networks. 1991;**4**:251-257

[27] Richard MD, Lippmann R. Neural network classifiers estimate Bayesian a posteriori probabilities. Neural Computation. 1991;**3**:461-483

[28] Carpenter GA, Grossberg S. A massively parallel architecture for a self-organizing neural pattern recognition machine, computer vision, graphics. Journal of Image Processing. 1987;**37**:54-115

[29] MacQeen JB. Some methods for classification and analysis of multivariate observations. In: Proceedings of the Fifth Berkeley Symposium on Mathematical Statistics and Probability. Berkeley: University of California Press; 1967. pp. 281-297

[30] Tran TA. A study of the energy efficiency management for bulk carriers considering navigation environmental impacts. Journal of Intelligent & Fuzzy Systems. 2018:1-14. DOI: 10.3233/JIFS-171698

Chapter 3

Households' Energy Efficiency Practices in a Bereft Power Supply Economy of Nigeria

Ibrahim Udale Hussaini

Abstract

The study focuses on attaining energy efficiency practices in the housing sector of the Nigerian economy. This is essentially necessary in order to reduce the energy demand on the central power supply of the nation and as well attain energy security. Nigeria as a nation is endowed with enormous energy resources, yet beleaguered with chronic energy crisis because of inadequate power supply to the citizens. The overall goal is to seek ways of improving the energy use situation of the country; and the objectives are to determine the prevailing levels of energy efficiency practices in housing design; appliances in use; and occupant behavior. The findings reveal a low level of energy efficiency consideration in housing design practice; a very low level of appliances efficiency; and a much low level of energy efficiency practice by the housing occupants. Thus, a strategic scheme of energy efficiency practices to be realized by the government and housing stakeholders is proffered for the housing sector of Nigeria.

Keywords: households, energy efficiency practice, bereft power supply, renewable energy resources, national economy, Nigeria

1. Introduction

The power drive of any nation is the state of energy supply which is the thrust of its national development in the many facets of its economy. The power supply should be adequate and sustainable in order to achieve a progressive and sustainable national development. In most situations, inadequate utilization of reliable and sustainable energy resources is the bane to the attainment of a sustainable power supply. But the case of Nigeria is paradoxical as the abundance of these resources does not portend adequate and reliable power supply.

However, the issue of energy has become one of the most sensitive discourses of our time; and as a result, the world is starting to accept the possibility of change in the patterns of consumption, leading to energy conservation measures and more rational use of existing energy sources to ensure sustainability. This change in perception is no more apparent than in the growing recognition that energy is the key to the development of the global civilization and essential to improving the quality of life beyond the basic activities necessary for survival.

According to United Nations Publication [1] energy use is keenly related to economic development, poverty reduction and the provision of vital services.

Nevertheless, its production, distribution and consumption can have adverse effects on global environment at either the local or regional levels. Consequent upon this realization, the contemporary society is faced with the challenges of developing technologies to improve access to modern energy services, increase energy efficiency and reduce air pollution; and initiating policies on energy consumption to meet future global energy demands with renewable resources. Thus, the need to adopt all possible measures to ensure that buildings use of energy is minimal i.e. Heating, Ventilation, Air conditioning and Cooling (HVAC); and Lighting systems are to use methods and products that conserve energy or reduce energy use. Furthermore, it is acknowledged that the technology-based improvement on energy efficiency is significantly influenced by the human social behavior in the utilization of the energy. In fact, well known energy analysts like Gerald Gardner, Lutzenhiser and Paul Stern have opined that a significant boost in a more efficient use of energy resources can be attained through understanding and shaping of human behavior [2].

The energy need of the society is rising daily and the pressure of sustaining this rising demand is becoming critical. Of interest is the energy consumption per household in developing countries which would be growing as income rises and more electrical appliances are installed thereby exerting rising demand on the central power supply. Therefore, to ensure sustainability in the built environment, there is the urgent need for developing countries like Nigeria to imbibe the policy of energy efficiency practice in the National Development Programmes which for now is absent or inactive.

At present, there is a prevailing state of apathy in the energy sector in Nigeria, particularly in the area of housing, with the accompanying energy inefficient households in all parts of the country. Hence, the needs for households' energy use reform through appropriate frameworks of energy efficiency practice necessary for sustainable development. To further understand the implications of the proposed reform, it becomes necessary to elucidate on the tripartite issues arising from household energy use pattern.

2. Tripartite issues of households' energy use

Formidable attempts at addressing issues of efficiency in household energy use should focus on housing design practice (architectural), the efficiency of appliances in use (technology), as well as the housing occupant behavior in the consumption of the energy as thus presented.

2.1 The architectural (design) issue

Housing is the shelter component of human existence, where he lives and sustains his worldly activities. Due to the current trends in civilization, housing design has embodied multiple considerations among which is the energy use pattern. This makes the issue of end-use energy in the built environment and particularly in housing more crucial than ever. Consequently, the responsibility of developing sustainable management scheme toward enhancing the quality of our environment through environmental and energy-conscious planning and design is saddled on the stakeholder personnel involved in the built environment. This obligation arises from the quest for better efficiency in the use of energy and other resources in our built environment. The result of this could be a new scope of architecture and construction, so that this branch of the industry can supply the contribution necessary for sustainable and viable development in reducing energy use, contrary to earlier assumptions that high energy consumption is reminiscent or suggestive of a superior culture. As such, the desired energy efficiency as relates to buildings should begin with the planning and design through construction to occupancy in consideration of the natural environment. However, the most cost-effective energy reduction in a building usually occurs during the design process which makes it crucial to review some aspects of the architectural technology in terms of design and services/appliances provision in the built environment because without technology and technological advancement, the tools we need to attain efficiency would not be available [3, 42].

2.2 The efficiency (technology) issue

Technology in this extent is the application of practical sciences to industry; and the efficiency component is the ratio of useful work done by a device to the energy supplied to it usually expressed as a percentage. Undoubtedly, our living conditions have being rapidly improving over the years since the era of Industrial Revolution in the fourteenth century. This is because suitable technologies and appliances, networks and synergies have been developed, resulting in more and more successes; and advances in all spheres of life being recorded on a continuous basis in the bid to meet our basic needs.

The quest to meet our basic needs in most times has resulted in several other problems impacting on the environment. These associated problems are initiated from the processes of industrial development, urbanization and resources exploitation in the form of environmental pollution and resources depletion; and are usually counter 'cost-effective.' To apprehend these adverse effects is the call for sustainable development which according to Hegger et al. [4] is to be accomplished by *Effectiveness* and *Efficiency* in the management of resources. They argued that the aspect of effectiveness refers to 'doing the right things' at whatever expense while the efficiency outlook which is 'doing things right' induces resources utilization to the barest minimum [42].

Applying efficiency measures can be low-cost or can require a significant investment, and could involve a conservation practice. The accompanying conservation practice refers to change of behavior in order to save energy (and money); e.g. turning off the lights when not in need. Nonetheless, both energy conservation and efficiency measures help one to reduce energy use, energy bills, air pollution and greenhouse gas emissions. Therefore, instituting energy efficiency practice through policies implementation is specifically due for third worlds like Nigeria where the energy demand is currently on the increase as households increase their appliances and equipment with improvement on their economic and social status whilst the national energy and central power supply is in a deplorable condition.

2.3 The behavioral (human) issue

The forces behind behavior in action are a complex phenomenon to discern and as such deserve a meticulous attention to unveil. Energy consumption in the home is such activity that requires the appropriate behavior to attain efficient utilization for the desired end-use. On this premise, it is assumed that energy consumption in the housing sector would be significantly influenced by behavior of the people as the basic users of the energy [5].

It is notably argued that the issue of human dimension to energy use cannot be undermined since consumption is still a poorly understood phenomenon, and simply because the variables that determine consumption have not been clearly identified [6]. As such, it becomes indispensable to have an understanding of the social and behavioral issues of our built environment so that many of the benefits of greater technological efficiency that would be attained may not be lost.

Residential building sector in Nigeria is the highest energy consumption sector of the economy, and is associated with energy efficiency problems. It accounts for about 50% or more of annual electricity power consumption [7] with an associated problem of wastage due to lack of energy saving measures in place. At the moment, the energy sector of the nation is undergoing power deficit as it lacks enough/ adequate power and energy to sustain her growing economy; prompting the need for increase in power generation and the institution of EE practice [8, 9]. Generally, there is a specific problem of higher energy use demand in the building sector, particularly in the urban areas arising from the rapid growing population, increase in living standards and rising number of apartments. This phenomenon calls for concern with immediate attention.

Apparently, the issue of energy has become a prime agenda of civilized nations of the world in recent times because of the circumstances surrounding its sustainability. Although, it is acknowledged that energy is the key factor to societal development in many spheres of life (economic, social and industrial), its existence is being threatened by the global fear of scarcity and high price. More so, that energy resources exploitation is being considered as one of the main causes of climate change and environmental damage is aggravating the situation. Therefore, the challenge of the sustainability of energy through efficient utilization and adoption of conservation measures to reduce the effects of the currently associated problems becomes paramount. Thus, the study is to proffer ways of improving the peoples' understanding of energy consumption and subsequently enhancing energy efficiency practice in the built environment.

3. Theories of energy efficiency and human behavior

3.1 Energy efficiency in housing

Energy efficiency is a phenomenal term which is technology focused; but imbibes a behavioral essence in practice [10]. Davidson and Henderson [11] presents energy efficiency as an indicator of the economic value obtained from the consumption of fuel, which when applied to housing, is best assessed in terms of the cost of energy needed to produce a given output or level of service such as a standard of heating. On this premise, they went further to define an energy-efficient house as one which when compared with houses of similar size, costs less to heat, to light, and to operate its essential services.

According to Ahsan [12], well-designed energy-efficient buildings remain the best environment for human habitation while minimizing the cost of energy consumed; with the objective of improving the comfort level of occupants and reduced energy use for heating, cooling and lighting. Janssen [13] considers this improvement in energy efficiency as any action undertaken by a producer or user of energy products, which decreases energy use per unit of output without affecting the level of service provided. This therefore signifies that an energy efficient house has good thermal insulation, efficient heating and lighting systems and probably, well-conditioned occupant behavior.

In the view of Majumdar [14] most environmental problems of today are related to buildings construction, occupancy and demolition due to their excessive consumption of energy and other natural resources; and the associated environmental pollution. Thus, the built environment is witnessing gross resources depletion and environmental damage arising from the imposed pressure by accelerated urbanization and the sought 'energy-intensive' solutions to our basic needs. Alleviating these problems require that we design and develop future buildings on a sound concepts of energy efficiency and sustainability. This could be accomplished by applying environmental/climate conscious design principles together with other multifarious approaches like the use of materials with low embodied energy, effective use of renewable energy resources, conditioning occupant behavior, etc.

It is pertinent to understand that more than one third of the world's energy is used in buildings; and a majority of that energy is particularly used in houses and apartments. One can therefore help humanity and save a lot of money by building a super-efficient house which uses only 10–30% as much energy as a house of similar size that is built to contemporary standards [15].

The potential benefits of energy efficient designs are immense. Of principal importance are the Europe-wide energy benefits following uptake of the climate-sensitive design. In northern Europe, passive solar gain and daylighting reduce the need for heating and lighting energy. In the United Kingdom, studies on passive solar housing have indicated a significant energy save of about 5% from improved site layout. Curiously, enormous energy could be saved by the application of sound concepts of sustainability in new buildings; and applying retrofit options to existing ones with an accompanied reduction in environmental pollution [16, 17].

Therefore, energy systems designed to be efficient, decentralized, and diversified are what national security demands, the public wants, and the market is ready to supply [18]. This can be achieved in the Nigerian households through collective efforts of the government and housing stakeholders in addressing the identified study issues.

3.1.1 Environmental and economic benefits in the delivery of energy efficiency

The cumulative (environmental and economic) benefit in delivering energy efficient buildings is in the accomplishment of reduced running costs, reduced environmental impact, improved ambient conditions and increased equipment life.

Therefore, creating buildings that use less energy not only reduces and stabilizes costs, but also reduces environmental impact. It is a fact that the knowledge and technologies to reduce energy use in our homes and workplaces without compromising comfort and esthetics is available now. But the prevailing situation is that the society is not taking full advantage of these advances because buildings are typically designed and operated without considering all the environmental impacts. These buildings can improve the health, comfort and productivity of occupants in measurable ways [19].

According to Littlefair et al. [16] cities are growing rapidly, and are increasingly polluted and have become uncomfortable places to be. Industrialization, the concentrated activities of dwellers and the rapid increase in motor traffic are the main contributors to increase in energy consumption and air pollution, and deteriorating environment and climatic quality. They contend that, the urban heat island effect generated can cause temperature differences of up to 5–15°C between a European city center and its surrounding, resulting in increase demand for cooling energy; and the increase in temperature may also exacerbate pollution by accelerating the production of photochemical smog.

Consequently, new developments are unfolding the world over in the uptake of the climate-sensitive, energy efficient designs to reduce excessive energy demands on the economy, and also to counter the increasing adverse effects of these developments by maximizing use of renewable energy sources and reducing energy dependence on fossil fuel, thereby minimizing carbon dioxide emission into the environment. In addition, the housing sector in Nigeria can achieve reduction in energy demand by directing effort on occupant behavior in household energy use.

According to Horsley et al. [20], one of the most significant environmental impacts of buildings occurs through the consumption of energy during their operational lives. And that, the effective management of the design process is pivotal in the delivery of buildings with improved efficiency but, unfortunately, the monitoring of energy performance is not currently a typical part of the construction design process; which in fact, is vital to be addressed.

In due consideration of energy efficiency standards for the society, it should be noted that the built environment has significant inertia, and in order to deliver a significant improvement in energy performance for the built sector, both new and existing buildings must be considered for assessment. Consequently, a culture of energy conservation will have to be fostered among all members of the project delivery chain, from clients to architects and contractors to building users before any significant improvements in performance will be noted. With this, the building industry will be able to deliver greater, more durable buildings with reduced whole life costs. It will also make a very significant contribution to reducing CO_2 emissions as a step toward a more environmentally acceptable way of living [20].

Although environmental reasons are strong, in practice cost savings usually drive energy efficiency. Therefore, energy efficiency measures should generally be considered in their order of economic payback, complexity and ease of application. Measures according to CIBSE guide [21] fall into three broad types:

- no-cost/low-cost requiring no investment appraisal,

- medium cost requiring only a simple payback calculation, and

- high capital cost measures requiring detailed design and a full investment appraisal.

It is however arguable that energy-efficient buildings do not actually cost more to establish than conventional buildings do. This is because the application of 'sustainability' and 'energy efficiency' concepts does present opportunities to offset or minimize costs of avoidable mechanical systems and services [21].

3.1.2 Energy efficiency design principles

Energy efficient building designs are credited partly to the adoption of climate and environmentally-conscious design principles by the creation of reduced energy loads in buildings. According to Majumdar [14], architects can achieve energy efficiency in building designs by studying the macro- and micro-climate of the site, applying bioclimatic architectural principles to combat the adverse conditions, and taking advantage of the desirable conditions. Subsequently, some common design elements have been identified to directly or indirectly affect the thermal comfort and visual conditions of building occupants, and thereby the energy consumption of buildings. Some of these elements according to literatures [12, 14, 22–24] as indicated below are the basis for the design of the 'housing evaluation form' (the checklist) for the case study:

a. *Planning/design consideration* [the building site, building typology/planform, building orientation, functional distribution (room orientation), landscaping, and the design process]

b. *Building envelope* [external walls and finishes, fenestrations and shading, thermal insulation, roof]

c. *Other services* [building materials, electrical and lighting installation, air conditioning installation]

3.2 Human behavior and energy use in the households; a theoretical framework

The third research issue (others being design and technology) is the behavioral issue which addresses the human dimension to energy use in the households. Thus, the theoretical background here presented provides the theoretical framework for the research survey (a quantitative approach) on household energy use analysis.

In fact, the idea of the 'behavioral approach' to energy use analysis according to Wortmann and Schuster [25] evolved from the apparently insufficient contributions of economy- or technology-based models to advise politicians on how to initiate developments toward energy conservation. The compelling scenario has made consumption a poorly understood phenomenon, as the variables that determine consumption have not been clearly defined [6]. Much technology-based improvements on energy efficiency have been accomplished but often dampened by the inappropriate human social behavior in the utilization of the energy [26]. In this regard, Diez-Nicolas [9, 27] present 'social ecosystem' and 'center periphery' theories to explain human attitudes and behaviors as relates to his actions. The former elucidates on attitudes as 'instrumental collective responses that a population develops under a given state of arts (technology) in order to achieve the best adaption to the environment;' while the latter unfolds that new attitudes are first developed at the center of the society before spreading toward the social periphery. He further argues that the concern for the built environment is more in developed societies; and among individuals of higher social status that are better informed [28].

Williams et al. [29] contend on the need to focus more on the behavior of the consumer in our attempt to conserve or utilize energy efficiently. Thus, attention should be given to how man uses his environment and how he metamorphoses in response to economic forces around him. Wilhite et al. [30] argue that energy use in the home is related to physical and structural variables like the dwelling's envelop, size, and appliances; and also to occupant behavior. But the behavior component is frequently underestimated or ignored in demand-side management (DSM) programs partly because of its complexity. The argument goes further to stress that human behavior is influenced by some interacting variables of socio-cultural traditions (attitudes, esthetic norms, comfort, symbols); economic considerations and *knowledge levels*. Based on this understanding, Beeldmann and Bais [31] have acknowledged the importance of knowledge about human behavior which they say is essential for successful energy savings policy.

In addition, Ehrhardt-Martinez [2] presents an argument that *effective policies* can make inconvenient behaviors convenient, and can as well make expensive behaviors less expensive. Instituting effective policies can remove structural, institutional, and legal barriers to behavioral change. In fact, understanding and shaping behaviors can provide a significant boost in the more efficient use of all energy resources. Nonetheless, the inefficient pattern of human behavior [32, 33] represents a large, untapped reserve that could potentially reduce current levels of energy consumption by 20–25%; and do so in ways that save money [2].

According to Golubchikov [34], the housing sector is one of the priority areas with regard to energy efficiency not only because it consumes a great amount of energy (up to 50% of total consumption in individual European states in recent years), but also because it remains *remarkably wasteful*. This is because the housing sector still and actually maintains outdated technology with inefficient practices, despite the high potential of the current existing technology to drastically reduce energy use in housing.

As a remedial measure, Ehrhardt-Martinez [2] has suggested that efforts to understand human behavior must start with the recognition that people are motivated to action as a result of both economic and non-economic factors. On this basis, it becomes necessary to have a strategic scheme (as proffered in this study) that would identify the individual as a rational economic and socio-political actor making rational choices regarding the adoption of more or less efficient technologies and behaviors. Meanwhile, Beeldmann and Bais [31] have identified two kinds of human behavior with respect to energy use;

    i. Investment behavior; which is related to the process of buying new appliances, equipment, goods etc., with a probable consideration of product efficiency. On this platform, questions that relate to factors that influence the buying of a product and why customers buy specific products are answered. This is usually related to moments of purchase. This can be influenced, monitored and measured.

    ii. User behavior; which relates to the actual use of products after the moment they are bought. It is concerned with questions of how often a product is used and in what way it is used. User behavior is important for energy consumption during the lifetime of the appliance. Influencing user behavior can have very large effects on energy savings, but of a fact, it is more difficult to influence, monitor and measure.

In another dimension, Sanquist [35] suggests 'curtailment' and 'efficiency' actions as the two principal types of actions necessary for energy consumption reduction. The former involves actual reduction in the frequency or duration of specific activities, such as single-car commuting; while the latter involves one-time actions, such as installing improved home insulation or purchasing new-model, energy efficient appliances. On the whole, curtailment involves repeated activity that produces relatively smaller energy reductions, while one-time efficiency actions involve greater expense that produce relatively larger energy reductions. Although, applying any of these dimensions, either investment/user behaviors or curtailment/efficiency actions, depends on the level of awareness of the individual—particularly on the knowledge of the implications/benefits of the actions to be undertaken.

In all, these theories have formed the basis for the derivation of the study variables (dependent—efficiency practice/practical behavior, and independents—education, awareness, and social status) and the drafting of the questionnaires on household energy use survey. Also, the energy use framework to be proffered is in consideration of these theoretical factors and the actual results of the survey, the interview and the case study.

To understand the situation better, a theoretical analysis of how attitude can be responsible for personal behavior becomes necessary.

3.2.1 Attitude and behavior; a theoretical analysis

In household energy use analysis, it is believed that attitude shapes behavior, but it is behavior that ultimately affects energy use [10]. On this background, social scientists see attitude as a predictor for behavior. In the past, they viewed attitude as individual mental process that determines a person's actual and potential responses; and as such, developed theories that suggested "Attitudes could explain human actions." In 1929, Thurstone, L.L. developed methods for measuring attitudes using interval scales; while in 1947, Doob adopted the idea of Thurstone that attitude is not directly related to behavior but it can tell us something about the overall pattern of behavior [36].

Consequent upon these developments, Ajzen and Fishbein [36] assume that individuals are usually quite rational and make systematic use of information available to them. And that people consider the implications of their actions before they decide to engage or not to engage in a given behavior. This proposition was referred to as the 'theory of reasoned action.' Though, it was later realized that this theory was inadequate and had several limitations [37] particularly with people who have little or feel they have little power over their behaviors and attitudes. The sequence of this development led from an earlier theory of reasoned action to the 'theory of planned behavior.'

In reality, humans can exhibit total control in certain behaviors if there are no attached constraints of any sort, but in situations where adopting a behavior requires the possession of a resource or skill which is absent in the individual; then a total lack of control is evident. On that account, planned behavior may embody control factors, which according to Ajzen and Fishbein [36] are either internal or external. Internal factors like skills, abilities, information, emotions such as stress etc.; and external factors may include such things as situation or environmental factors. Thus, the postulations that the individual's intention to perform a behavior are a combination of attitude toward performing the behavior and subjective norm. The subjective norm being the influence of social pressure that is perceived by the individual to perform or not to perform a certain behavior; while the attitude toward the behavior includes; behavioral beliefs, evaluation of behavioral outcome, subjective norm, normative beliefs, and the motivation to comply [36]. In fact, the subjective norm can be influenced to some degree by policy formulation as recommended in the proffered strategic scheme.

Meanwhile, attitudes and subjective norm are measured on scales like the Likert scale, using words or phrases such as like/unlike, good/bad; agree/disagree, satisfactory/unsatisfactory etc.; while the intent to perform a behavior depends on the product of the measures of attitude and subjective norm. Nevertheless, the individuals are more disposed (i.e. intend) to engage in behaviors that are believed to be achievable [38]. Hence, the adoption of this scale (Likert scale) as the factor of measurement of practical behavior (energy efficiency practice) in this research survey.

3.3 Energy and the Nigerian economy

Nigeria is the most populous country in black Africa (over 160 million people) with a very high abundance of natural resources, but with a very poor, weak and slowly improving economy that is heavily dependent on the oil sector. Although, Nigeria is one of the world's largest oil producing countries, it is currently experiencing rampant energy poverty due largely to the inefficiency of the energy industry to meet the energy demands of its customers. Electrical infrastructure is extremely scarce in Nigeria. Only 40% of Nigerians have access to electricity. Although more than 70% of the population lives in rural areas, only 10% are connected to the grid. Nigeria faces a serious energy crisis due to declining electricity generation from domestic power plants. Power outages are frequent and the power sector operates well below its estimated capacity. Often without prior warning, the average Nigeria firm is without power for over 15 h a week. For this reason in the year 2000, 2400 MW of electricity was being generated by diesel and petrol generating sets (EPIC, 2004 in Odularu and Okonkwo [39]) to run households and some other sectors of the economy. To ameliorate the situation, the government then claimed to be able to create infrastructure so that up to 85% of the population has electricity by 2010 which at the moment in 2018 remains a farce.

The residential housing sector remains the dominant area of electricity consumption among others like industrial and commercial [7]. In fact, the household energy consumption constitutes a substantial amount of societal energy demand resulting from rapid growing population, increase in living standards and the rising number of apartments.

Though, the electricity demand in the country is growing faster than the country's population, the electricity per capita is one of the lowest in the world [40, 41]. Nigeria per capita power consumption is estimated some time ago at 82 KW when that of South Africa is put at 3793 KW [39].

The utilization of renewable energy sources in Nigeria remains quite limited, although there is a realization that the renewable energy sector must grow in order for the country to develop sustainably. Solar power is being promoted as a method to improve electricity service to rural villages not connected to the national grid but in a very slow and negligible pace. There are renewed efforts by NGOs and the Centre for Renewable Energy Development in Nigeria (CREDN), urging the government to boost the use of renewable energy sources to diversify the country's energy consumption from petroleum.

Solar photovoltaic (PV) is an attractive method to try because it offers modularity and requires no fuel, but very basic and relatively simple operation and maintenance. It has long lifetimes with very little performance degradation which makes it much more suitable for rural environments and private individuals' exploration; yet this potential remains highly untapped in Nigeria.

4. Approaches and methods

The study is in three parts to address the objectives based on the issues of housing design, appliances in use, and human behavior in the utilization of household energy using the mixed-method approach.

On the issue of human behavior in household energy use, a theoretical background has provided the dependent variable of human behavior (this time; level of efficiency practice) with the independent variables of education, awareness, and social status of the individuals as the basic parameters which are subjected to a quantitative study approach. Here, the study identifies the households' respondent group as the target population which is defined as "all heads/representatives (adults) of household units (male or female) resident in Bauchi-Nigeria, and living in formal residential housing typology of flats/apartments, and not in traditional settings within the time period of this research." The sampling procedure for the group (household residents) is 'cluster sampling' of selected residential housing neighborhoods that are part of the target population. This is because the sampling frame for the entire housing units in Bauchi town could not be established. Cluster sampling therefore, allows for random selection of population elements in clusters, in which case, a multistage (i.e. two-stage) or clustering procedure was applied to identify clusters (groups of housing units) and then sampled within them. As a result, six distinct clusters of housing units were established in six different locations of the three geographical districts of Bauchi Local Government Area (i.e. 2 clusters per district) for the survey so as to be able to generalize the outcome of our results on the entire target population.

Both quantitative and qualitative study approaches were adopted in the determination of the energy efficiency consideration in housing design practice in Bauchi town. For the quantitative approach, the population for this group (professional practice respondents) is defined as "all housing stakeholders (male or female) in the building industry (architects, building service engineers and builders) resident in Bauchi-Nigeria within the time period of this research." The sampling procedure was purposive (judgmental) and the population was determined from the register of professional associations of the respective disciplines in Bauchi town. The sample size was the entire population due to their meager sizes, and more so that not every member of the population would respond due to some unavoidable factors of availability etc., and in some cases, outright refusal to participate in the survey. However, the qualitative approach was through case study of selected housing units (12 nos.) from the created study clusters (housing neighborhoods) within Bauchi town using a checklist (evaluation form) of energy efficiency design variables as derived from literature. Also, an interview strategy of selected and validly determined housing stakeholders (professionals in practice) in Bauchi town was undertaken using 'structured—interview questions.'

At another level, the determination of energy efficiency level of households' appliances and lighting is undertaken using a qualitative approach by taking inventory of appliances and lighting in selected housing units (12 nos.) from the study clusters within the three geographical districts (Bauchi, Galanbi and Zungur) of Bauchi (metropolis) LGA. This was achieved by use of checklist based on the approved/certified appliances' energy efficiency rating/labelling as available in literature as well as the building market.

In the overall analysis, the qualitative data was subjected to content analysis while the quantitative data was subjected to both descriptive and inferential statistical analyses to obtain results.

5. Results and discussions

The result of the survey conducted on energy efficiency in the residential neighborhoods of Bauchi, Nigeria elucidates on the issues of design practice, appliances in use and human behavior (focus of study) in EE practices.

On the behavioral aspect of the study, the independent variables of education and awareness; and the social status of residents have provided indicators on the levels of perception and practice of energy efficiency by the people. This indeed will foster opportunities for appropriate policy and regulations. The levels of education, awareness and social status of individuals were quite above average (greater than 50%) and that is a good recipe for energy efficiency, though actually determined by practice. On the contrary, the subsequent result on the EE practice in all the study clusters is quite unimpressive (less than 50%). To have a good record of EE practice, the rating (score) of a majority of households must be in the range of 5–7 on the Thurstone scale and should be well above 50%. But in this case, it is just 48.8%. The majority rating in the range of 1–4 on the scale (51.2%) as shown in **Figure 1** indicates gross inefficiency practice. The situation portrays household energy efficiency practice in Bauchi, Nigeria to be on a 'much low level' in spite of the higher levels of education and social status. Probable factors like life style and culture; the lack of awareness and absence of appropriate polices on energy issues are responsible for this result.

The correlation analyses of the relationship of the dependent variable (practical behavior) with the independent variables of education, awareness and social status have indicated varying levels of significance. This implies that these independent variables may have somewhat degree of influence on the practical behavior of the individual in household energy use but may not necessarily determine it as in this case [3, 42].

Household Energy Efficiency Practice Rating (%level)

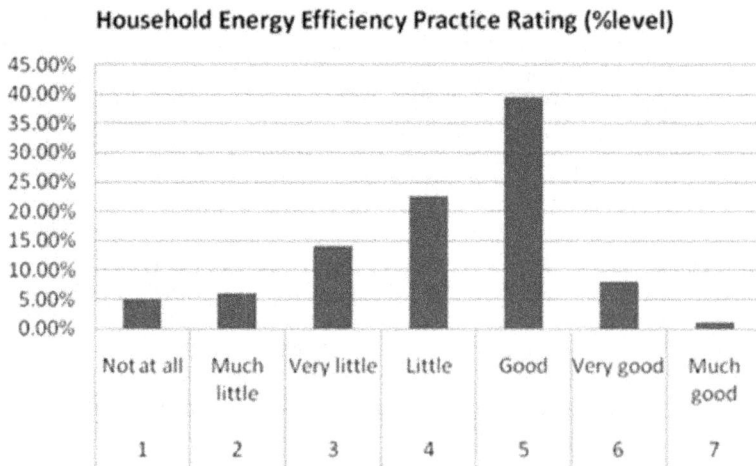

Figure 1.
Rating of household energy efficiency practice on Thurstone scale.

Data on housing design practice was obtained through survey research using both quantitative and the qualitative approaches. Although, there was a satisfactory level of 'concern for energy conservation' from the questionnaires administered and interviews conducted on practitioners, their respective levels of awareness was unimpressive with a dampened effect of EE consideration in design practice. However, the result of the inventory (case study) indicates some varying levels of 'adequacies' in design variables (at low level) and 'inadequacies' (at high level) in energy efficiency/conservation considerations in design practice. This assessment is done in consideration of the design elements of building typology, building orientation, the building paved area, window size/openability, daylighting, cross ventilation, plan form, functional distribution, shading from trees/structures, placing of windows against solar radiation, plant landscaping, ratio of built form to open spaces, incorporation of water bodies, shading devices, etc. Although, these elements do not seem to have direct influence on energy efficiency of a building, they help to facilitate energy load reduction on buildings. They constitute factors of consideration in climate/environmental conscious design. Thus, the cumulative result of the qualitative study (interview and case study) on the level of energy efficiency consideration in housing design practice is a 'low level' one (indicating less than 30% for interview and 54% for case study); and also 'low level' for the quantitative study respectively. Several factors are responsible for this low result among which are absence of appropriate policies by the government and the lack of guidelines to regulate practice in the direction of efficiency by the professional bodies.

The third issue of consideration is the energy efficiency level of appliances and lighting in use in the households. An inventory of some selected housing units was undertaken using a checklist based on appliances EE description as ordinary (not efficient), efficient type, and undetermined (no identified efficiency rating/labelling). The lighting appliances were categorized into incandescent and fluorescent; and the fluorescent further grouped as ordinary type and efficient type (CFLs) as available in our society. The result indicates that only 36.7% of lighting points in the surveyed households are energy efficient. Acknowledging the fact that lighting is a fundamental aspect of energy consumption in Nigeria as it is the dominating appliance in use in all of the households. Therefore, promoting efficiency in lighting alone can lead to immediate enormous gains. In short, lighting and all appliances of cooling, heating, cooking, refrigeration and electronics are in the category of low and very low scores on a 5-point scale respectively. These scores fall below the desirable 'high' to 'very high' scores necessary for the accomplishment of good EE practice in the households' appliances. Hence, the cumulative result indicates a 'very low' level of energy efficiency in household appliances in Bauchi, Nigeria. Furthermore, the finding has also indicated that cooking in Bauchi, typical of Nigeria and other underdeveloped countries (Africa) has both the modern and traditional methods incorporated in a majority of the households. The modern ways include the use of electric cooker, electric stove, gas cooker, micro-wave oven, etc., while the traditional methods include the use of kerosene stove and wood as fuel. The latter is a substitute and an alternative to the former in most cases because of the scarcity and high price of the electric energy and cooking gas required to utilize these appliances. Meanwhile, kerosene and wood are relatively cheap and readily available. As a result, almost all households subscribe to this substitute. However, it is discovered that among the electricity-type appliances only 10% are energy efficient. Therefore, improving the reliability of power supply along with the use of energy efficient appliances would assist to remedy the energy crisis situation in Nigeria.

6. The energy challenges and the way forward to attaining household energy efficiency

Nigeria as a nation has being battling with energy crisis for decades in the form of inadequate power supply and inefficient utilization of the end-use energy. To combat the crisis, persistent efforts are being made by the government, the private sector, NGOs, etc. to ameliorate the situation; yet there still persists the inability to overcome the energy poverty of the nation. This is due to many factors among which is the inadequate funding of energy projects, ardent corruption in administration setting, and the lack of adequate information/know-how on energy matters in terms of sustainability of resources and end-use efficiency. In addition, there are several other barriers identified to particularly hinder the mainstreaming of energy efficient appliances in Nigeria; e.g. policy barrier, legal and regulatory barrier, technical barrier, research and development barrier, etc. [43].

To generally tackle the energy situation there is the need to explore increased penetration of renewables into the energy supply mix [7]; and to particularly expend significant effort in addressing efficiency issues of design, technology and behavior in the housing sector. The energy sector has successively been engulfed with fierce corruption that requires a strong will and logical determination of the government to stamp out through the implementation of a strategic plan.

Furthermore, the global quest for sustainable development in the environmental, social and economic dimensions coupled with the demand to attain energy security has prompted a dire need for a strategic scheme of energy efficiency practice for a bereft energy economy of Nigeria. More so that most nations of the world have instituted energy efficiency programs in this direction, Nigeria with a chronic energy crisis cannot be left behind. Examples are; the energy efficiency strategy of South Africa launched in April, 2004; the US Energy Commission on Behavioral and Social Sciences in 1985 and several individual states programs like the Texas ENERGY STAR Home Programme, Guarantee Home Programme; EU Energy Commission, etc. [42].

6.1 Main pillars of the strategic action plan for energy efficiency practice and its attendant obstacles

There is no gainsaying the fact that the current and persistent energy poverty in Nigeria requires the implementation of a strategic plan or scheme to recover. Hussaini and Abdul Majid [42] have expounded much on this plan in the aspects of policy formulation and implementation, research and development, public information and participation, technology/housing evaluation and monitoring, financial incentives and market introduction, and institutional strategies.

Effective policy formulation and implementation is the beacon of genuine and progressive national development; and this should be fashioned to regulate energy use in the direction of efficiency. This is necessary in order to enforce control in the manner and pattern of energy consumption in general. Emphasis should be laid with details on (i) design practice which is to be safeguarded with professional practice codes; (ii) technology procurement and its marketability; and (iii) the attainment of conditioned human behavior in the utilization of the energy in recognition of Ehrhardt-Martinez [10] postulations that 'effective policies do in fact make inconvenient behaviors convenient; and expensive behaviors less expensive.'

The government and the various stakeholders in the building industry should endeavor to promote research and development in all aspects of the built environment in which end-use energy is operatively involved. Research constitutes the primary tool used in all fields of endeavor to expand the frontiers of knowledge. It is the key factor to societal growth and development. This is because, progress made in every aspect of life depends on the contributions made by systematic research. Therefore, research into architectural design practice should embrace the aspect of energy use in the home which according to Wilhite et al. [30] is related the physical and structural variables of the buildings; and should also include both technology deployment in appliances manufacture and occupant behavior in order to determine which aspect requires behavioral attention or demand side management (DSM) attention in our complex energy use phenomenon. Funding avenues for research and development directed at promoting design and technology innovations in conjunction with occupant behavior on energy efficiency should be identified and adequately established.

The need for adequate and reliable information on energy efficiency matters available to the people, and their comprehensive/inclusive participation in energy efficiency programs for a successful outcome is unequivocal. This is because the power of public information is undaunted, and the effect of people-oriented participation is sublime. On this note, Laitner et al. [33] has suggested the establishment of a 'people-centered initiative' to promote public participation in energy savings in both active and passive ramifications.

Technology as the back bone of societal development is the application of practical sciences to industry and commerce which in the building industry should be monitored and evaluated for performance and possible improvement both in appliances installation and housing design. Technology procurement is necessary for achieving desired innovation, while the monitoring and evaluation exercises are put in place in order to determine the level and pattern of consumption so as to identify the gray areas of needed improvement in attaining concrete energy targets.

Motivation of stakeholders and the public to be actively involved in energy efficiency matters can be achieved through financial incentives and market introduction. Indeed, Laitner et al. [33] have argued that great efficiency gains can be achieved through financial incentives and motivation to the public in the form of subsidies on energy-efficient products and services. However, the introduction of market transformation is to arouse the supply of energy-efficient products and services by technology procurement and should be provided in the arena of all-inclusive and all-embracing residential energy services.

The vision and missions of a culture of efficiency practice in the housing sector can be accomplished through organizational strategic planning. Establishing institutional strategies is the realm of the professional stakeholders and should effectively steer the practice procedure in the direction of efficiency with ethical connotations. Continuous Practice Development (CPD) programs should be fervently entrenched in professional practice so that stakeholder practitioners are adequately trained to always imbibe efficiency practices and to be kept abreast in new trends and developments.

6.2 The obstacles

Hussaini and Abdul Majid [42] have lucidly outlined the hitches in our household energy-use phenomenon that are likely to be responsible for the absence or lack of structures in the main features of the strategic scheme as follows:

1. the lack of comprehensive National Energy Policy resulting from certain barriers of policy, legal and regulatory origin;

2. financial constraints due to low budgetary measures on energy efficiency matters;

3. technical incapability due to lack of adequate experts in the area of energy efficiency;

4. the low level of public awareness due to "lack of the willingness" of the government and housing stakeholders to adequately mobilize and sensitize the public toward energy efficiency; and

5. the persistent widespread corruption of the Nigerian public in all aspects of the society (private and public) where all that matter is the immediate material gain out of little or no significant effort.

Regardless of the above obstacles, a strong and keen determination of the government and all parties concerned to implement the strategic plan for EE practice as indicated is a good recipe to overcoming any impediments on our energy efficiency path.

7. Recommendations

The under-given recommendations have been offered in recognition of the tripartite issues and the eventual research findings.

The first recommendation is to strictly adhere to the strategic scheme of EE practice as specified in order to realize the goal and objectives.

However, the primary concern in addressing the issues should be focused on energy saving, including all the possible methods to accomplishing it; and should form the policy hallmark at all levels of decision making. Thus, the idea of energy saving should be promoted throughout all structures of the society including the family and at all levels in schools and educational systems. This should include the public and the professional communities (stakeholders in housing provision), explaining the main challenges and what can be done to save energy and reduce greenhouse effect. This may require the provision of energy manuals to housing residents accompanied by periodic household energy use monitoring exercise.

In summary, the study suggests a three-way (tripartite) practical approach to achieving energy-efficient households, and as well improving on energy efficiency practice based on some researchers postulations (as derived from literatures) in the aspects of building design (architectural), services/appliances design (technology) and conditioning occupant behavior (behavioral).

7.1 Improving on the architectural (design) dimension

The issue of energy efficiency practice in residential buildings cannot be ignored, and should be accomplished by the following measures;

- Reduce energy consumption by improvement on building design and thermal isolation according to modern climate control principles.

- Qualified energy audit of buildings should be carried out before executing energy saving measures.

- Stimulating research, development and demonstration of modern technologies/design techniques and their use within the context of domestic energy resources and local conditions.

7.2 Improving on the technology (appliances/services efficiency) dimension

- Reduce energy consumption by substitution of electric heating by other energy sources.

- Reduction of energy consumption of air conditioning systems and lighting. Lighting in particular takes up to 20% of total electric energy produced worldwide.

- Reduction of energy consumption of electronic equipment in stand-by mode.

- Development of windows with heat transfer coefficient less than 2 W/m^2 K and the development of special glasses with increased reflection and selective absorption and emission abilities.

- Consideration of the passive energy design strategies for natural lighting (daylight) and natural ventilation; and the development of measurement techniques for evaluating the energy efficiency of buildings.

- Development of ventilation and air conditioning systems based on heat and moisture recuperation is particularly important.

- Setting energy efficiency standards by the controlling authority by imposing minimum level of efficiency though manufacturers are not usually disposed to it because of administrative and adaptation costs.

- Stimulating the supply of efficient products by technology procurement, i.e. offering incentives to manufacturers to take part in development and diffusion of highly energy efficient products.

- Introducing DSM (demand-side-management) at the appropriate levels where environmental condition is sufficient as a control.

7.3 Improving on the behavioral (human) dimension

- Saving of energy in households by stimulation and support measures which include, introducing new methods of energy consumption management, educating people through the electronic and print media; and giving financial incentives in the form of subsidies in the price of energy efficient materials.

- Stimulate consumer choice by labeling of appliances to enable consumers compare efficiency of certain products; and showing annual power consumption which translates into running costs.

8. Conclusion

It is evident that Nigeria is witnessing increased population growth which is associated with increased energy demand and consumption in all facets of the economy. The increased energy demand and the prevailing inefficient pattern of consumption coupled with the associated environmental issues in energy delivery system has been a cause for concern. Therefore, the current call for energy efficiency practice is not out of place, and could play a valuable role in guiding the society in the choice of the energy path to follow.

The proposed schematic plans or strategy for the realization of energy efficiency practice is justified because the residential building sector for now is energy-inefficient and also the largest energy demand sector of the Nigerian economy. This is provided as a wake-up call to the government, the housing stakeholders as well as the public for the attainment of energy efficiency. It prescribes the application of sound concepts of sustainability and energy efficiency through the deployment of environmental/climate-sensitive design principles together with multifarious approaches like the use of materials with low embodied energy, effective use of renewable energy resources and conditioning occupant behavior. It is also essential to disabuse the mindset that energy-efficient building costs much more to establish than a conventional building. It does in fact offsets avoidable costs due to installation of energy-intensive mechanical systems and services.

Interestingly, the implementation of the strategic scheme would eventually lead to sustainable environments of a free or declined environmental damage/pollution where energy conservation is utmost, producing healthier and more viable productive settings.

Declaration

This is to acknowledge the fact that this work constitutes majorly on the excerpts from the following works by the same author [3, 42].

Author details

Ibrahim Udale Hussaini
Department of Architecture, Abubakar Tafawa Balewa University Bauchi, Nigeria

*Address all correspondence to: hudalib@yahoo.co.uk

IntechOpen

References

[1] United Nations Publication. Trends in Sustainable Development. New York: Department of Economic and Social Affairs (DESA), Division for Sustainable Development; 2006. ISBN: 92-1-104559-2. [Accessed: 20-06-2007]

[2] Ehrhardt-Martinez K. Behaviour, Energy, and Climate Change: Policy Directions, programme Innovations, and Research Paths. Report Number E087. American Council for an Energy-Efficient Economy (aceee). 2008. http://www.aceee.org

[3] Hussaini IU. Household energy efficiency practice in Bauchi, Nigeria [unpublished PhD thesis]. Department of Architecture, Kulliyyah of Architecture and Environmental Design, International Islamic University Malaysia; 2012

[4] Hegger M, Fuchs M, Stark T, Zeumer M. Energy Manual: Sustainable Architecture. Munich: Birkhauser; 2008

[5] Lutzenhiser L. Social and behavioral aspects of energy use. Annual Review of Energy and Environment. 1993;**18**:247-289

[6] Wilk R. Towards a Useful Multigenic Theory of Consumption. Conference Proceedings of European Council for an Energy Efficient Economy (eceee) 1999 Summer Study-Panel 3.15: Human Dimensions. 1999. http://www.eceee.org/conference-proceedings/ [Assessed 16/04/2010]

[7] Sambo AS. Renewable Energy Development in Nigeria. In: A Paper Presented at the World Future Council/Strategy Workshop on Renewable Energy; 21-24 June, 2010; Accra, Ghana. 2010

[8] WADE. More or Less: How Decentralized Energy Can Deliver Cleaner, Cheaper and More Efficient Energy in Nigeria; A Report by World Alliance for Decentralized Energy (WADE), Christian Aid and International Centre for Environment and Energy Development (ICEED); July, 2009. www.localpower.org

[9] IAEA. Energy Indicators for Sustainable Development: Guidelines and Methodologies. Vienna: International Atomic Energy Agency; United Nations Dept. of Economic & Social Affairs, International Energy Agency, Eurostat & European Environment Agency, IAEA; 2005

[10] NEB. Attitude and Behaviour Shaping Energy Use. Alberta Canada: National Energy Board (NEB), Energy Briefing Note, The Publications Office; November 2009. pp. 1. ISSN: 1917-506X

[11] Davidson PJ, Henderson G. Improving Energy Efficiency in Housing. BRE Information Paper (IP 24/89). Garston, Watford WD2 7JR, UK: Building Research Establishment, Department of Environment; 1989

[12] Ahsan T. Passive design features for energy-efficient residential buildings in tropical climates: The context of Dhaka, Bangladesh [unpublished M.Sc. thesis]. Kungliga Tekniska Hogskolan, Stockholm: KTH Department of Urban Planning and Environment. Division of Environmental Strategies research-fms; 2009. www.infra.kth.se/fms [Assessed 29/03/2010]

[13] Janssen R. Towards Energy Efficient Buildings in Europe. London: The European Alliance of Companies for Energy Efficiency in Buildings; 2004

[14] Majumdar M editor. Energy-Efficient Buildings in India. New Delhi, India: Tata Energy Research Institute, Darbari Seth Block, Habitat Place, New Delhi & Ministry of Non-Conventional Energy Resources; 2002

[15] Wulfinghoff DR. How to Build & Operate a Super-Efficient House. Version 040118. Wheaton, Maryland USA: Wulfinghoff Energy services, Inc.; 2003. p. 1

[16] Littlefair PJ et al. Environmental Site Layout Planning: Solar Access, Microclimate and Passive Cooling in Urban Areas. London: BRE Publications; 2000

[17] Majumdar M. Energy Efficiency in Green Buildings—An Integrated Approach to Building Design (Green Business Directory; CII-Godrej GBC). GBC

[18] Lovins AB, Lovins LH. Energy Forever. American Prospect. 2002;**13**(3):30-34

[19] Torcellini P. Better Buildings by Design. Solar Today. Boulder USA: American Solar Energy Society; March/April 2001. pp. 40-43

[20] Horsley A et al. Delivering energy efficient buildings; a design procedure to demonstrate environmental and economic benefits. Journal of Construction Management and Economics. 2003;**21**:345-356

[21] CIBSE Guide. Energy Efficiency in Buildings. London: Chartered Institution of Building Services Engineers; 1998

[22] Watson D, Labs K. Climatic Building Design: Energy-Efficient Building Principles and Practice. New York: McGraw-Hill; 1983

[23] Mallick FH. Thermal comfort and building design in the tropical climates. Energy and Buildings. 1996;**23**(196):161-167

[24] Givoni B. Climate Considerations in Building and Urban Design. New York: Van Nostrand Reinhold; 1998

[25] Wortmann K, Schuster K. The Behavioural approach to energy conservation: An opportunity still not taken by energy policy. In: Conference Proceedings of European Council for an Energy Efficient Economy (eceee) 1999 Summer Study-Panel 3.17: Human Dimensions. 1999. http://www.eceee. org/conference-proceedings/ [Accessed: 16-04-2010]

[26] Bell M, Lowe R, Roberts P. Energy Efficiency in Housing. England: Avebury Ashgate Publishing Limited; 1996. p. 87

[27] Diez-Nicholas J. Measuring and explaining environmental behaviour: The case of Spain. In: Dooley B, editor. Energy and Culture; Perspectives on the Power to Work. England: Ashgate; 2006. pp. 209-229

[28] Galtung J. Foreign policy opinion as a function of social position, in Diez-Nicholas J. (2006). Measuring and Explaining Environmental Behaviour: The Case of Spain. In: Dooley B editor. Energy and Culture; Perspectives on the Power to Work. England: Ashgate; 1964. pp. 209-229

[29] Williams DI, Crawshaw AJE, Crawshaw CM. Energy efficiency and the domestic consumer. The Journal of Interdisciplinary Economics. 1985;**1**:19-27

[30] Wilhite H, Nakagami H, Masuda T, Yamaga Y, Hanada H. A cross-cultural analysis of household energy-use behaviour in Japan & Norway. In: Conference Proceedings of European Council for an Energy Efficient Economy (eceee) 1995 Summer Study-Panel 4: Human Dimensions. 1995. http://www.eceee.org/conference-proceedings/ [Accessed: 16-04-2010]

[31] Beeldman M, Bais JM. Modeling human behaviour for policy decisions. In: Conference Proceedings of European Council for an Energy Efficient

Economy (eceee) 1994 Summer Study-Panel 4: Human Dimensions. 1994. http://www.eceee.org/conference-proceedings/ [Assessed: 16/04/2010]

[32] Gardner GT, Stern PS. The short list: The most effective actions U.S. households can take to curb climate change. Environment. 2008;**50**(5): 12-24

[33] Laitner JA, "Skip", Ehrhardt-Martinez K, McKinney V. Examining the Scale of the Behaviour Energy Efficiency Continuum. Forthcoming. Washington DC: American Council for an Energy-Efficient Economy; 2009

[34] Golubchikov O. Green Homes: Towards Energy-Efficient Housing in the United Nations Economic Commission for Europe Region. ECE/HBP/159. New York & Geneva: United Nations; 2009

[35] Sanquist TF. Human factors and energy use. Human Factors and Ergonomics Society (HFES) Bulletin. 2008;**51**:11

[36] Ajzen A, Fishbein M. Understanding Attitudes and Predicting Social Behaviour. Englewood Cliffs, NJ: Prentice-Hall; 1980

[37] Godin G, Kok G. The theory of planned behaviour: A review of its applications to health-related behaviours. American Journal of Health Promotion. 1996;**11**(2):87-98

[38] Bandura A. Self-efficacy: The Exercise of Control. New York: W. H. Freeman & Co.; 1997

[39] Odularu GO, Okonkwo C. Does energy consumption contribute to economic performance? Empirical evidence from Nigeria. Journal of Economics and International Finance. 2009;**1**(2):44-58

[40] Garba B. Demand side management and efficient lighting initiatives in Nigeria. In: Paper Presented at the World Energy Council (WEC), Africa Workshop on Energy Efficiency; 29th June, 2009; Addis Ababa, Ethiopia. 2009 [Assessed 10/2/2011]

[41] CIA World Fact Book. 2012. Available from: www.world.bymap.org/ElectricityProduction.html; www.photins.com/rankings/economy/electricity_consumption_percapita_2012_0.html [Accessed: 13-04-2012]

[42] Hussaini IU, Abdul Majid NH. Energy development in Nigeria and the need for strategic energy efficiency practice scheme for the residential building sector. Management of Environmental Quality: An International Journal. 2015;**26**(1):21-36

[43] UNDP—Nigeria EE Appliances Project Document. Promoting Energy Efficiency in Residential and Public Sector in Nigeria; for UNDP Supported GEF Funded Projects. 2010. Available from: www.TheGEF.org [Accessed: 13-04-2013]

Chapter 4

Frontiers of Adaptive Design, Synthetic Biology and Growing Skins for Ephemeral Hybrid Structures

Sandra Giulia Linnea Persiani and Alessandra Battisti

Abstract

The history of membranes is one of adaptation, from the development in living organisms to man-made versions, with a great variety of uses in temporary design: clothing, building, packaging, etc. Being versatile and simple to integrate, membranes have a strong sustainability potential, through an essential use of material resources and multifunctional design, representing one of the purest cases where "design follows function." The introduction of new engineered materials and techniques, combined with a growing interest for Nature-inspired technologies are progressively merging man-made artifacts and biological processes with a high potential for innovation. This chapter introduces, through a number of examples, the broad variety of hybrid membranes in the contest of experimental Design, Art and Architecture, categorized following two different stages of biology-inspired approach with the aim of identifying potential developments. Biomimicry, is founded on the adoption of practices from nature in architecture though imitation: solutions are observed on a morphological, structural or procedural level and copied to design everything from nanoscale materials to building technologies. Synthetic biology relies on hybrid procedures mixing natural and synthetic materials and processes.

Keywords: adaptive design, membrane technology, synthetic biology, ephemeral design, sustainable design

1. Introduction

Manipulation of the environment can arguably be considered as a natural trait of adaptation in a broad range of animals, from nesting and building of complex architectures to the use of tools in mammals, birds, reptiles, fish and some invertebrate species. Mankind remains however the undisputed leader in the field, and membrane structures encompasses a big share of the early tools employed by Man. First made of natural skins, then woven fabrics and as technology evolved, progressively more and more synthetic materials have been employed to manufacture membranes for wearables, packaging and shelters. In fact the end-use of membrane structures has not drastically changed since. The features and the complexity of the manufacturing however have.

IntechOpen

Membranes are traditionally classified into synthetic or biological, and are in both cases essential for life on (and outside) Earth, being responsible for regulating all type of energetic exchange between a given organism and the synergetic system(s) it is part of. The nature of each membrane varies with its function, and can differ fundamentally in structure, size, transparence, etc. [1]. Today, in the age of nanotechnology and gene manipulation, technology prepares for a new paradigm shift where the borders between natural and artificial, designed and evolved, produced and grown become ever more indistinct.

Technological innovation and scientific intuition are strongly influenced by other fields, among others Design and Art, as (r)evolution in one domain impacts the others [2] and developing markets can powerfully drive innovation. As technology and science rediscover how performing Nature-evolved solutions actually are, and how important it is for us to design sustainably, preserving the balance of a system we are a part of, adaptivity becomes an interdisciplinary rising business and trend. Automated homeostasis and transient features to integrate in artificial artifacts become sought-after aspects even in Architecture, a very conservative sector, where design has for a long time been interpreted as in distinction or even in opposition to Nature. Innovations in materials and technologies are very rarely developed in this field: solutions are traditionally built to last for long times and are applied over very big scales, hence prioritizing cautious and low-cost solutions. Introducing change is risky and needs to be justified by consistently adding efficiency to the system. This new rising model is therefore bringing a true revolution to the whole sector: it involves on one hand an intellectual effort to rethink dogmatic preconceptions as longevity, stability, and performance in built environments, and on the other, the incorporation and adaptation of new technologies and materials [3]. Generally, new concepts and materials are first adapted in more progressive and experimental fields that are closely related to Architecture, as Art and Design, which dare to take bigger risks.

The field of Arts and Entertainment is wealthy and free enough from building codes, users' needs and requirements to allow experimentation. Being consistently smaller, artworks are generally far less expensive to manufacture than buildings, allowing more experimentation and a broader diffusion by being displaced and exhibited in nonstandard locations to reach a broader public. Art is therefore a great occasion to test and advertise ideas, raising the interest of users and developers. It is not a chance that many new solutions that have further developed in architecture have started as part of an artwork or an exhibition pavilion.

Industrial design is today going through huge changes due to the growing interest and demanding taste of consumers, the use of new materials and technologies, which allow the insertion of the most charming features, opening up to new dimensions of esthetically choreographing change. As expressed by designer Raymond Loewy, "Ugliness does not sell," and companies commit a lot of attention in designing every aspect of a product. Today a huge innovation potential is linked to new materials, which can develop entirely new concepts and markets: products become animated, adding character, life and desirability [4].

This chapter introduces, through a number of examples, the broad variety of hybrid membranes in the context of experimental Design, Art and Architecture. The case-studies are categorized following two different stages of biology-inspired approaches. The first, Biomimicry, is founded on the adoption of practices from nature in architecture though imitation. Solutions are observed on a morphological, structural and procedural level, then copied to design everything from nanoscale materials to building technologies. The second approach, Synthetic Biology, relies on hybrid methods mixing natural and synthetic materials and processes.

In order to enable to overcome old preconceptions and widen the conceptual boundaries, "Membranes" will in this context be defined in the more primitive sense of the word, *a thin selective barrier allowing some effects to pass, while halting others*.

2. Biomimetic, inspiration and imitation of nature

Biomimicry introduces the concept of observing, understanding nature, and learn from the fittest solutions instead of searching what to extract from nature or harvest parts of organisms as raw material. Ideas are borrowed and reinterpreted into another context leaving Nature untouched and available for others to draw inspiration from [5]. Inspiration often proceeds from biology to design, as a natural phenomenon suggests a new way of solving a challenge, but the process can also be inverted, from design to biology, where a challenge in the technical world is identified and a solution is searched for among organisms or ecosystems achieving similar functions.

These solutions are still very new to the market and, for a great majority, too expensive for being used in architecture. Useful applications can however be found in the near future, for temporary shelters, pneumatic membrane structures, adaptive facades as well as for multifunctional and responsive interiors. Our buildings are evolving towards a non-mechanic, material-integrated adaptivity allowing the structures to meet external and internal changes in climate and user behavior.

2.1 Biomimetic material structures

With the emerging field of nanomaterial technologies, scientists become the architects of matter. Materials are designed with unique proprieties observed in natural materials, "hacked" and artificially designed for man-made applications.

2.1.1 Membranes with enhanced performances

We are rediscovering how Nature proportionally outperforms synthetic man-made structures in almost all aspects: spider silk is 10 times more resistant than Kevlar [6], grass stalks are thinner and more flexible than any man-made bridge, etc. The observation and study of these structures on microscopic level allows a more thorough understanding of how Natures does it, and to copy the solutions. Thanks to nanotechnology, a whole new category of engineered materials is being manufactured with boosted proprieties: super resistant, ~absorbing, ~light, etc.

Spacer textiles, or 3D textiles, have a dual wall structure with a space in between of a few millimeters up to tens of centimeters, giving the possibility to create a specific microclimate in the cavity, making the material light and robust. These are used in a broad range of contexts, from functional clothing, furniture and vehicle design, construction and transport of temporary structures [7].

Drag reducing fabrics are highly flexible and light spandex/nylon composites with drag-reducing water-repellent features, mimicking the surface of shark fins, which have made them the new high-tech innovation in the field of competitive water sports, improving glide through water with a 38% reduction in resistance. Manufactured swimsuits are designed covering arms and legs, with bonded seams to further reduce drag, also providing compression to maximize muscle performance and reduce the entry of water between the suit and the body [8].

Soft organic photovoltaic cells (Soft PV) are a new rising technology in the field of Building Integrated Photovoltaics (BIPV). Researchers are testing printed polymeric PV cells fabricated onto carbon nanotube-based electrodes upon various flexible and translucent ethylene tetrafluoroethylene (ETFE) building components [9]. The prototypes are thought of as possible prefabricated shading systems, easy to integrate in energetic retrofittings of existing structures.

2.1.2 Propriety changing membranes

In imitation of many defensive adaptations in nature, a growing category of smart materials is being developed with propriety changing features. These materials change one or more of their characters in reaction to influencing factors as light, heat, humidity, etc., and have in the last years been emerging in functional design and clothing, not only at a conceptual stage, but are in some cases market ready.

Self-healing membranes are still a young research area, with an anticipated enormous economic and sustainability potential. These materials, polymers and elastomers in the case of membranes, autonomously counter degradation and micro-damage by adding a repairing agent or acting from inside, eventually in response to an external stimulus [10]. German and Swiss researchers are working on a biomimetic liana-plant inspired solution to realize a self-healing polymer membrane for load-carrying pneumatic structures for lightweight constructions. The principle is the expansion of a two-component polyurethane and polyester foam, as a temporary "first aid" layer autonomously expanding in the event of a hole in the pneumatic structure and a sudden exposure to a rise in pressure [11, 12].

Thermo-regulating textiles using micro-encapsulated Phase-Change Materials (PCM) are relatively new to the textile market, although they have been employed for about 30 years by Nasa. These textiles allow absorption, storage and release of thermal energy at pre-programmed temperatures, for widely diversified applications from sportswear to bed textiles [13].

The *"textile in a can"* patented technology allows to spray non-woven fibers (natural, synthetic, recycled or biodegradable), liquefied in an evaporating solvent, directly upon surfaces, allowing the fibers to bond forming an instant fabric. The material can be applied with varied degrees of hardness (even as casts for broken bones), is repairable (re-sprayable) and recyclable. Most of all, from a designer's point of view, this type of innovative application dispenses the realization process of any artifact from the constraints of cutting, stitching or fitting the surface to its support (**Figure 1**) [14].

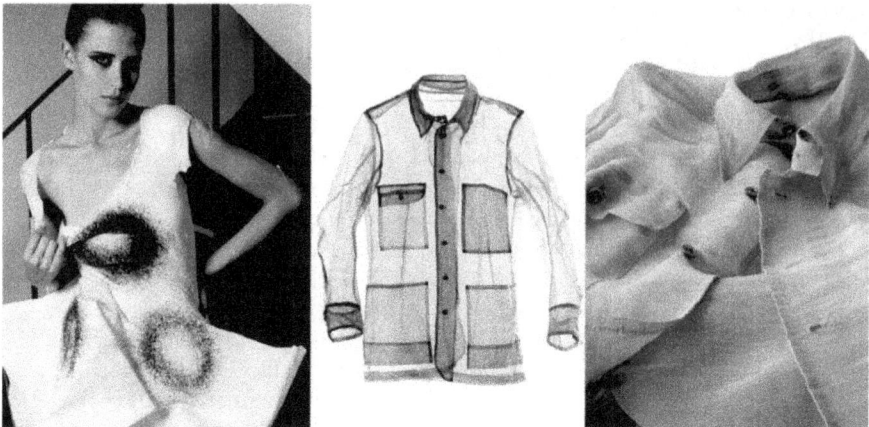

Figure 1.
Dress realized with the sprayable textile technology [14], and the "Oricalco" shape-memory shirt [15].

Shape memory (SM) fabrics integrate smart fibers (mostly SM polymers or Nitinol, a Titanium alloy), with the ability to recover a pre-programmed shape, in reaction to changes in temperature (or in some cases to light). Applications in the field of Design include both wearables and furnishings [16]. A small-medium Italian enterprise manufactured a long sleeved shirt using SM alloys, programming its "autonomous ironing" if heated up under a flux of hot air, as that of a hairdryer [15].

2.1.3 Ecological membranes

In the case of certain applications, using a material, which automatically breaks down and dissolves after a set time, becomes an asset. Following the example of Nature, which lives and thrives in the same ecosystem where it manufactures and recycles its own substances [5], a growing number of natural and biodegradable films and surfaces are being developed for packaging first and foremost, but not only. Moving past petroleum-based artificial fibers, towards protein-based building blocks, we could be on the verge of a new textile revolution. In this case, man uses Nature as a co-worker, developing new techniques to craft materials in a way that is more similar to gardening and farming than to manufacturing [17]. As decaying processes are not reversible, the applications in design and architecture become not only interesting, but also extremely innovative.

Natural fiber membranes are developed in a huge variety of raw materials, in many formats and for different uses, from fire- and tear-resistant banana paper [18], mushroom leather used for surfaces from shoes to furniture [19], cork composites available even in thin flexible sheets and bark cloth lampshades [7].

Bioplastics and *biocomposites* are, from being considered as niche products, quickly developing and expanding in importance. A team from Barcelona's Iaac (Institute for advanced architecture of Catalonia) has developed bioplastics from food waste based on orange peels [20]. In a similar direction, Dutch designers developed an algae-based polymeric bioplastic fit to dry and process into a 3D printable material (**Figure 2**) [21].

Engineered spider silk has been attempted by many material researchers, as spider silk is known to be one of nature's strongest materials. As spiders cannot be farmed, scientists and companies are attempting to mimic this natural protein-based fiber. Researchers at the University of Cambridge have designed non-toxic highly tensile-resistant hydrogel fibers made 98% out of water. Apart from the proprieties

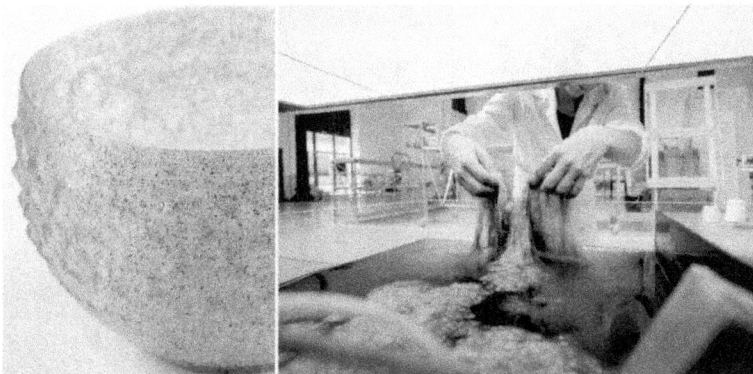

Figure 2.
3D printed cup with algae-based filaments (right), realized by Luma Foundation in collaboration with Musée Départemental Arles antique; (left) sample of Cladophora macroalgae [21].

mimicking those of the spider silk (although not nearly as strong), the new method has shown how synthetic fibers can be manufactured without relying on high-energy and toxic processes [22].

2.1.4 Interactive membranes

Interaction seems to be the new frontier for materials as well as in many other fields, in the Era of Informatics. Smart and interactive fabrics have enhanced virtual proprieties being enabled to sense and communicate information, taking us a few steps closer to Artificial Intelligent (AI) systems. Technology becomes wearable and integrated into all kinds of products from toys to life-saving devices.

Light emitting fibers are closing the technological efficiency gap with the organic LED (OLED) technology, offering however low-cost and low-energy manufacturing conditions. Light-emitting electrochemical cells (LECs), are generally single-layer devices sandwiched between two electrodes, area-emitting light in any color. These can be obtained from 100% environmentally friendly raw materials, promising cost-efficient applications that have so far been applied on or integrated with plastics, paper, textile, and metal [23]. Lightweight and flexible, the fibers can potentially be woven into textiles to create smart fabrics for any application from wearable electronics to next generation lighting.

Monitoring fabrics are still experimental and mostly rely on micro-electro-mechanical systems (MEMS) to measure physiological parameters for health monitoring and protection. This is the case of the stretchable ultrathin display consisting of micro LEDs mounted on a rubber sheet, designed by Japanese researchers to transmit wireless biometric data to a cloud platform [24]. A different case is the conductive cotton thread developed by researchers of the University of Michigan, "smartened" by infusing the fiber in carbon nanotubes and polymer solutions with added antibody anti-albumin able to detect blood, which has a potential use in high-risk professions [25].

Electroactive polymers (EAPs) make artificial muscles. "ShapeShift" is a dynamic surface material to explore the potential of its application in Architecture. The elements are made of pre-stretched films on flexible acrylic frames, sandwiched between two compliant electrodes, and able to stretch under the action of high DC voltage. Through the connection of more elements maximization of the kinetic effect was enabled, allowing the structure to support itself (**Figure 3**) [26].

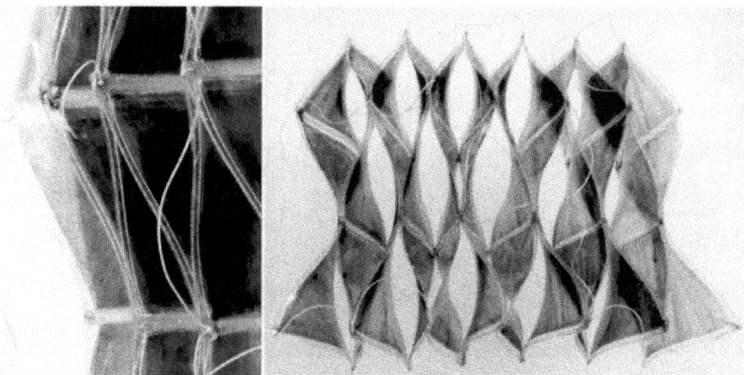

Figure 3.
Artificial muscle membrane "ShapeShift" [26].

2.2 Biomimetic design

Moving on from the microscopic scale of material design, to the scale of Industrial Design, Nature is used as a model of inspiration to craft man-made. For the future of architecture, the improved performances mean not only the chance to reinvent completely new aesthetics and cultural approach, as in every material revolution, but most of all it opens up to the possibility of imagining completely new, previously inexistent functions and uses.

2.2.1 Design with nature

Nature is in this case used as a partner. Organic materials are used fully or partially, and "crossbred" to create new solutions.

The *Edible water bottle* is a transparent spherical edible seaweed membrane designed by a British startup as an alternative to the petroleum-based plastic bottles that are producing huge amounts of waste. Looking like a giant water drop, which can be made in various sizes, the gelatinous capsule bursts under a light pressure delivering its content, which can also be used for soft drinks, spirits and cosmetics. The recipe is public and can be replicated by anyone who obtains the ingredients [27].

Concrete cloth is a 3D membrane structure combined with dry concrete and a waterproof PVC layer on one face. The cloth is first bent into the wished shape and then hydrated, allowing the concrete to harden and the fibers to reinforce it. Originally developed for erosion control and rapidly deployable shelters, it has been used by a number of designers pushing its strength and flexibility as far as possible, achieving light structures in total contrast to what is normally associated with concrete (**Figure 4**) [28].

Wooden textile is a hybrid combination of two kinds of natural materials aimed at conveying new sensory experiences. Half wood and half textile, the wooden tiles laser-cut and stuck to one side of a fabric, transform the material into a structured but soft and flowing surface. Flexibility, mobility and weight depend on the size and thickness of the combined wooden tiles [30].

Figure 4.
The Whorl Console made with the concrete cloth technology [29].

2.2.2 Imitation of nature

Specific characters that we recognize as features of living organisms are imitated, adding not only functionality but also beauty. The references to organisms and animal features become an integral part of the concept: although the features are abstracted, the achievement becomes all the more successful the more the plagiarism is evident.

BMW Gina is a concept car with a groundbreaking design and an external flexible skin in polyurethane-coated Lycra [7], an extremely durable, flexible, water-repellent textile fabric stretched across a movable metal wire skeleton. Functions are revealed when needed through the translucent material or moving the substructure, giving access to the service points in the engine [31]. The membrane imitates the mechanics and features of a natural skin, strengthening the association of the machine to a living animal.

The Moving Mesh is a project for an adaptive sun shading façade able to withstand wind loads thanks to its folded surface geometry. The prototype is realized in an aluminum composite sheet about 3 mm thick, enclosing a highly elastic material, which creates flexible hinges when exposed through milling, allowing over 80,000 damage-free bending cycles. The shading element works as a single perforated surface, inspired by the flexibility and porosity of the human skin. The geometry drives the opening and closure of the diamond shaped flaps in the same way as an origami surface (**Figure 5**) [32].

Tape Paris is a temporary installation displaying a stretched biomorphic skin made out of transparent packaging tape, forming 50 m long hollow passageways suspended at a height of 6 m from the ground. The "parasitical structure" is supple and elastic, revealing its interior visitors through its translucent surfaces [33].

The *Louis Vuitton Matsuya Ginza Facade*, realized with aluminum sheets coated with a pearlised fluoropolymer paint, is an imitation halfway between a natural skin and a textile, repeating an art-deco pattern as a reference to the brand [34].

Mushtari is a one-piece sculpture printed by MIT researchers in a combination of plastic materials with different transparence and density. Imitating the shape of human interiors, the sculpture's 58 m hollow tubes are filled with a bacterial luminescent liquid in view of combining future versions with organisms capable of photosynthesis. The idea is for this wearable energy generator is to allow interplanetary travel and survival [35].

Figure 5.
"The Moving Mesh" prototype in scale 1:10 for a shading element [32].

2.3 Biomimetic processes

When imitating natural processes, it is not as much the final shape or the structure of an organism that is the focus of the analysis, but more time-related features as its creation and successive transformations. From an architectural perspective, research focuses on the potential of introducing behavioral patterns with envelope - and structural adaptivity on one hand, and on innovative production processes on the other.

2.3.1 Imitation of movement

Hylozoic Ground is an interactive sculpture environment installed within the Canadian Pavilion at the 2010 Architecture Biennale in Venice. It embodies a forest of suspended dynamic geotextile (acrylic) structures responding to its surroundings: a flexible transparent meshwork skeleton with ribbed vaults and basket-like stem allowing it to stretch, swell and bend composes the suspended artificial plants. The skeleton is made of partly flexible core parts and long rigid roller chain arms moved by Shape-memory alloy (SMA) wire rods. Pulling each rod, these tendons produce an upward curling motion lifting the latex membranes at the end of them in the air [36].

Hypermembrane is a standardized self-supporting structural system of adaptable shape and size designed for temporary to long-lasting lightweight architectures. The flexibility of the system is based on industrialized thermoplastic flexible structural components, which can be assembled in a three-dimensional interwoven structural mesh. The assemblage process allows not only adaptation of shape, but also reuse for different purposes [37].

In the *Experiment Cyclebowl* pavilion, the facade is made out of a series of three-layered Texlon foil cushions. A positive/negative leaf pattern is printed on the outer two layers of the system. The middle cushion is movable by changing the pneumatic pressure between the layers. By overlapping the pattern with the outer inversely printed cushion it transforms the facade from translucent to opaque to darken the interiors for the presentation of the shows inside the pavilion. Also, by assuming intermediate positions, variable ranges of shading are possible ranging from 45% light to complete darkness [38].

HygroScope: Meteorosensitive Morphology is a humidity-sensitive and responsive installation. The parametrically designed surface is built out of a multitude of maple veneer and synthetic composite triangles programmed to react differently depending on fiber direction, length, thickness and geometry. Absorption of water particles causes the distance between the fibers of the wood to increase, resulting in a swelling or lengthening of the material in the direction of the fibers: the composite triangular flaps curl, opening the surface's geometry. As the humidity rate drops, the panels reversibly straighten out closing the surface again (**Figure 6**) [39].

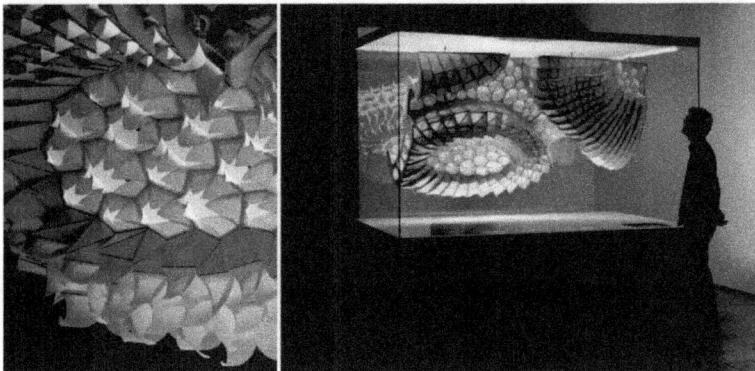

Figure 6.
HygroScope: Meteorosensitive Morphology at Centre Pompidou, Paris [39].

2.3.2 Digital fabrication

Digital fabrication, as a new tool for controlling additive design and manufacturing, is opening up an unprecedented potential to model and fabricate artifacts, realizing customized one-of solutions on industrial scale. In combination with parametric modeling and the introduction of new materials, 3D printing technologies open up the possibility to directly intervene and manipulate the structural and functional proprieties of artifacts. Features as size, geometry, translucency, elasticity and more can be closely shaped, functions are merged, and performances can be programmed all in one building block—as in natural structures, designed by evolution and built through biological growth.

Fluid Morphology is a translucent multifunctional 3D-printed façade element, developed at the Associate Professorship of Architectural Design and Building Envelope at TU Munich [40]. The research aimed to show the potential of additive manufacturing and 3D-printing technologies to close the digital chain loop in the industrial development of multifunctional building envelopes, from digital design and planning to the final product. The prototype is a one-piece rigid transparent polycarbonate shell; a façade element handling through its complex geometry heterogeneous façade functions as sun-shading, visual connection, acoustic deflection, load-bearing, insulation and ventilation (**Figure** 7).

The 3D printer allows the complex construction of a multiple compartmentalized element, interweaving façade functions, which are traditionally separated in layers in the contemporary building systems. This multiple functionality is not only material and energy saving, but also simplifies disassembly and recycling processes, reducing the need of separating technological parts made of multiple materials, with a consistent impact on the life-cycle of each single component, and by extension on the whole building.

In a similar way, the use of robotic construction technologies in architectural research is showing huge potential benefits over traditional construction methods in terms of speed, costs and complex custom-made geometries. The new methods of automatized distribution of material through the use of small semi-autonomous robotic agents open up new perspectives for the realization of on-site instantaneous and material-saving architectures.

The *Interactive Panorama* research pavilion developed at the University of Stuttgart [41] explores the potential of digital design and robotic construction applied to a bio-inspired method for pneumatic fabrication. The construction uses a

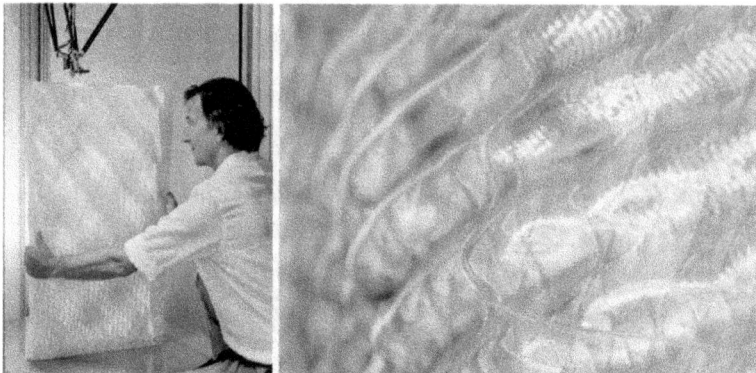

Figure 7.
"Fluid Morphology," 3D-Printed Functional Integrated Building Envelope built in a one piece transparent weather-resistant polycarbonate [40].

flexible pneumatic formwork inspired by diving bell water spider webs (*Argyroneta aquatica*). A robotic arm was placed inside a pneumatic ETFE envelope, which gradually stiffened by selectively applying layers of carbon fiber from the inside. The result is a thin rigid self-supporting carbon fiber composite shell.

Swarm Printing is an innovative approach to additive manufacturing technologies, where MIT researchers used live silkworms to grow a natural silk filament pavilion [42]. Robotic agents were used to rearrange the sticky and fast-growing filaments of 6,500 worms, following a hexagonal shell framework. As all worms were still available after the completion of the project, and potentially able to produce enough offspring to build another 250 pavilions, the process can be seen as a self-propagating material-producing colony.

3. Synthetic biology, hybridization of natural and artificial

In Synthetic biology (or *synbio*), Nature is no longer a simple inspirational model, but becomes the object of fabrication, as a new generation of Genetically Modified Organisms (GMOs). However, while genetic engineering is about copying, cutting and pasting DNA sequences, synbio rewrites and programs from scratch synthetic DNA to engender new applications, ultimately aiming to create "artificial life." Fostered by the engineering perspective on complexity, the practice reflects a highly systematized scientific approach to the understanding of biological procreation: biology is stripped to its bare bones, broken up into hierarchically abstracted parts (basic building blocks) that can be modeled using sequencing and fabricating, with the support of computer-aided-design (CAD) [43]. Unsurprisingly, the topic raises the same fears surrounding GMOs, concerning bioethics and security issues, as synthetized DNA is introduced in the food industry with self-replicating synthetic life forms [44]. As the limits between living and non-living blurs, questions are raised over where life begins, and how complex it must be.

In the field of Design and Architecture, the merging between biology, chemistry and nanotechnology is a farther-off reality where the hybrids are closer to the non-living than to the living. In an industry dominated by artificial petroleum-based products, the introduction of natural features and semi-natural organisms appear as less threatening, opening up to new ecological concepts and functional possibilities. Architects argue we are already in the Anthropocene, where it is no longer possible to distinguish where Nature begins and where it ends: we are part of a hybrid environment, a digital and rapidly urbanizing society where Mother Nature no longer exists and humans contaminate all ecosystems. The concept of "ecology" should therefore be revised and extended to embrace the biotechnological [3]. A synbio revolution in the construction field could lead to sustainable answers to our polluting and downcycling lifestyles, as factories are replaced by "biofactories," growing products with self-assembling, self-replicating, self-repairing, self-sustaining and self-degrading proprieties of living organisms.

3.1 Synbio materials

Synbio enables us to reconfigure living organisms, usually yeast or algae, to create man-made variants with pre-programmed features in order to perform specific tasks with a predicted outcome. These materials are still in their early stages of conceptualization and prototyping, but with great potential of future implementation also in architectural contexts, replacing the existing solutions with their biological highly efficient counterparts.

Protocells are basic non-living molecules that when stimulated by specific chemical cocktails can exhibit specific behaviors typical of living cells: response to pressure, light, heat, move, metabolize, reproduce, etc. The synthetic production of living material is therefore so far limited to basic applications, but has the potential to revolutionize the way we make materials and tools, blurring the gap between living and non-living.

A *synthetic bio leaf* is the first prototype of a silk protein-chloroplast based protein with photosynthetic capabilities developed by a London-based engineer [45]. As any natural leaf, this synthetic counterpart produces oxygen and absorbs CO_2 if provided daylight and water, with a huge potential of improving its efficiency with genetic modification. The energetic and environmental potential of this technology is huge if we think that light could be used as the ultimate energy source and carbon dioxide as the ultimate carbon source (**Figure 8**).

Engineered spider silk. A different attempt to create spider silk, other than the ones previously mentioned, has been attempted producing a protein-based synthetic silk through fermentation of GMO yeast, water and sugar. The ingredients are renewable and biodegradable, allowing to use cleaner manufacturing than the current technology [46, 47].

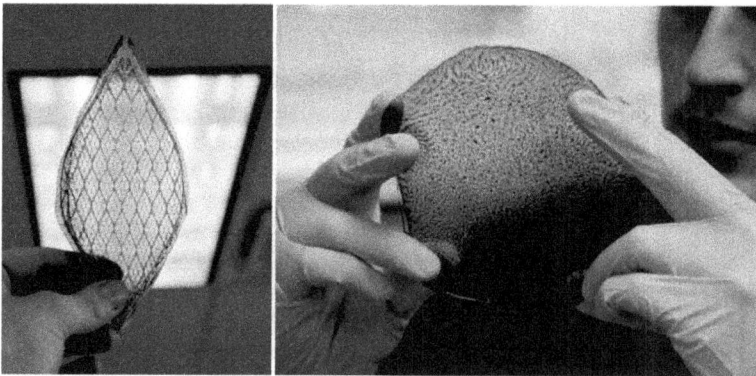

Figure 8.
"Silk leaf," the first biological membrane capable of photosynthesis [45].

3.2 Synbio design

Synthetic reprogrammed biological matter is introduced into Design and Architecture, envisioning tools and spaces with completely new, previously impossible functions, opening a window on a future of engineered living organisms. With synbio as a new design option, we face a new paradigm shift in the decades to come: how should we rethink our surroundings and our artifacts as they shift from mechanically dynamic to truly alive? In other words, becoming *semi-living tools*.

Amoeba running shoes is a speculative project for self-repairing shoe soles, based on a tailored protocell technology. These lab created non-living cells can be reprogrammed through chemical manipulation to acquire chosen abilities and behaviors of living cells: in this case inflation or deflation in response to pressure, adapting to the running substrate's texture providing cushioning. As the cells would be worn out after use, the application of a protocell liquid would allow the non-living organisms to regenerate [48].

Bioluminescent plant. Scientists have attempted a number of prototypal specimens of GMO light-emitting plants. A hale cress (*Arabidopsis thaliana*) has been provided genetic circuitry from fireflies [49], and fully functional bacterial luciferase pathways have been implanted in tobacco (*Nicotiana*) [50]. Although these biotechnologies are only at a prototypal level and scientists are working to improve the levels of light emission, the technology could revolutionize the world of lightning design.

3.3 Synbio processes

While biomimetic processes imitate natural procedures in artificial contexts, in Synbio the boundaries between the two have merged and closed the loop. The aim is to steer the natural processes towards prefixed goals. Most projects are however still at a concept stage.

Synbio fabric dyes are developed to manufacture environmentally friendly cost-effective long-lasting dyes entirely using genetically altered bacteria to fix dyes onto items of clothing. These innovative processes promise to consistently reduce water use and pollution of the conventional industrial dying processes [51, 52].

Synthetic biology and biology alike would greatly benefit from a deeper insight into the organized processes in cells, helping to understand how genes can amplify or inhibit its own expression [53]. Among the next frontiers of natural processes to imitate in synbio design:

Networks of metabolic reactions could theoretically be engineered from mammalian cells that have more complex structures. The idea would be to develop monitoring cell communities able to release therapeutic compounds [53]—in other words a new generation of monitoring substances.

Mutation control. Ideally, engineered designs should function for as long as possible, and neither crumble in the face of evolution, nor take unwanted paths. For that, microbial strains that are less susceptible to mutation can be used [53]—imagining in the future bio-artifacts with controlled aging processes. Our tools would not exhibit signs of age before their programmed end of use.

Reproduction of cells could revolutionize the way we manufacture, ideally controlling the timing of start and stop of reproduction, as well as the amounts. We can imagine this could be an alternative to healing materials, our facades and building surfaces autonomously replacing the broken and worn out parts with new material.

Programmed death of cells. As an imitation of the behavior of the lambda phage bacterial virus, which stays undetected for its host until it activates a program that ultimately kills the bacterium, engineers can use similar strategies to control cells not performing as engineered—or in the case of advanced materials and artifacts to program complex decay processes.

4. Conclusions

This chapter has reviewed cutting edge examples of membrane structures and materials used in the context of experimental Design, Art and Architecture. The categorization of membranes following a Biomimetic and a Synthetic biology approach has revealed, as could be expected, huge conceptual and technological differences between the two, and a great majority of case-studies that have been developed in the first category, while the second one is still at its very beginnings. These fields of study are the ultimate example of the effect of technological ephemeralization on our ability to do more and more with less and less.

From the analysis of this state-of-art, we can anticipate how there is a huge potential to introduce groundbreaking technological innovations, with both cultural and technical impacts on our everyday lives, and most of all on the environment.

From a functional perspective, more and more complex behaviors could be embedded in objects and components, ideally reducing the amount of tools and parts that we use. As our mobile phones today integrate multiple functions previously achieved by separate tools (phone, mail, camera, etc.), coming up with new uses and opening new markets, a rising integration of functions will be possible also in buildings.

From an aesthetic point of view, architects and designers need to accept change as a fundamental condition and rethink the nature of objects and environments. Time becomes an essential dimension to fashion, together with parameters as coordination and rhythm, which must be considered as central as function and form in our creations.

From a manufacturing perspective, materials might be grown sustainably and renewably, reducing polluting emissions on one hand and downsizing waste and over-production on the other, by growing materials only where and when these are necessary.

In the context of the building industry, there is the potential to overcome the constraints and the complex organization of traditional industrial and construction sites. By combining fabrication and construction in small-scale automatized solutions, large structures could be realized directly in urban areas, requiring the only presence of trained operators, the robots and any necessary material components.

Architects, designers and artists need to become increasingly part of these design processes, contributing with ideas and concepts, and most of all helping to shape a cultural and ethical approach to these topics. As sustainability, which is generally understood as a need for minimizing the impact of Man on the environment, the core of the question is to understand what we are sustaining, for whom and for how long. Technological advancements are transforming the world, but might not always do so towards the better—requiring a sound criticism towards the destined outcomes of an unleashed development. Solutions should not be predetermined but found pursuing a desired outcome.

Acknowledgements

The research has been funded by Ateneo Sapienza 2017; Scientific coordination Prof. A. Battisti, Operational coordination Ph.D. Sandra G. L. Persiani.

Conflict of interest

The authors declare that no conflict of interest exists, including any financial, material, personal or other relationship, which could influence the scientific work of this manuscript.

Frontiers of Adaptive Design, Synthetic Biology and Growing Skins for Ephemeral Hybrid...
DOI: *http://dx.doi.org/10.5772/intechopen.80867*

Author details

Sandra Giulia Linnea Persiani[1*] and Alessandra Battisti[2]

1 Chair of Building Technology and Climate Responsive Design, Department of Architecture, Technical University of Munich, Germany

2 Department PDTA, Faculty of Architecture, Sapienza University of Rome, Rome, Italy

*Address all correspondence to: sandra.persiani@yahoo.it

IntechOpen

References

[1] Mulder M. The Basic Principles of Membrane Technology. 2nd ed. Springer Science + Buisness Media, BV; 1996. DOI: 10.1007/978-94-017-0835-7

[2] Oxman N. The age of entanglement. Journal of Design and Science. 2016. Available from: http://jods.mitpress.mit.edu [Accessed: Mar 21, 2015]

[3] Kretzer M, Hovestadt L, editor. ALIVE: Advancements in Adaptive Architecture. Vol. 8. Applied Virtuality Book Series. Basel: Birkhauser Verlag GmbH; 2014. ISBN: 978-3-99043-667-7

[4] Ben Hopson, About Kinetic Design [Internet]. 2009. Available from: http://www.benhopson.com/?page_id=88 [Accessed: Feb 10, 2015]

[5] Benyus J. Biomimicry: Innovation Inspired by Nature. New York: Harper Perennial; 2002. ISBN: 9780061958922

[6] Agnarsson I, Kuntner M, Blackledge TA. Bioprospecting finds the toughest biological material: Extraordinary silk from a Giant riverine orb spider. PLoS One. 2010;5(9):e11234. DOI: 10.1371/journal.pone.0011234

[7] Peters S. Material Revolution, Sustainable and Multi-Purpose Materials for Design and Architecture. Basel: Birkhauser; 2011. ISBN: 978-3-0346-0663-9

[8] Tang SKY. The rocket swimsuit: Speedo's LZR racer. Science In The News (SITN), Harvard University, The Graduate School of Arts and Sciences [Internet]. 2008. Available from: http://sitn.hms.harvard.edu/flash/2008/issue47-2/ [Accessed: Apr 14, 2018]

[9] Zanelli A, Campioli A, Monticelli C, Beccarelli P, Hibraim HM, Maffei R. SOFT-PV: Creation of a Photovoltaic Organic Cell on Fluoropolymeric Substrate to Integrate into Smart Building Envelopes [Internet]. Politecnico di Milano: Textileshub; 2009. Available from: http://www.textilearchitecture.polimi.it/soft-pv.html [Accessed: Apr 19, 2018]

[10] Yang Y, Urban MW. Self-healing polymeric materials. Chemical Society Reviews. 2013;42(17):7446-7467. DOI: 10.1039/c3cs60109a

[11] Peter M. Self-healing membranes, nature shows the way. Empa Materials Science and Technology [Internet]. 2011. Available from: https://www.empa.ch/web/s604/nature-shows-the-way [Accessed: Apr 11, 2016]

[12] Rampf M, Speck O, Luschsinger RH. Self-repairing membranes for inflatable structures inspired by a rapid wound sealing process of climbing plants. Journal of Bionic Engineering. 2011;8(3):242-250. DOI: 10.1016/S1672-6529(11)60028-0

[13] Covered in Comfort. Report no. 20050031205. Nasa Technical Reports Server [Internet]. 2005. Available from: https://ntrs.nasa.gov/archive/nasa/casi.ntrs.nasa.gov/20050031205.pdf [Accessed: Apr 15, 2018]

[14] Fabrican. Fabric Form a Can [Internet]. 2018. Available from: http://www.fabricanltd.com [Accessed: Apr 10, 2018] (Photo credits: Fabrican Ltd, photographer Gen Kiegel)

[15] Grado Zero Innovation. Shape Memory Alloys [Internet]. 2017. Available from: https://www.gzinnovation.eu/material/7/shape-memory-materials [Accessed: Apr 11, 2018] (Photo credits: Grado Zero Innovation - Grado Zero Espace)

[16] Ritter A. Smart Materials in Architecture, Interior Architecture and

Design. Basel: Birkhauser; 2007. ISBN
978-3-7643-7327-6

[17] Collet C. En vie/Alive. Alive: New
design frontiers. Exhibition. Espace
Fondation EDF, Paris [Internet]. 2013.
Available from: http://thisisalive.com/
about/ [Accessed: Apr 14, 2018]

[18] Grado Zero Innovation. Pure
Vegan Banana Paper-Layer [Internet].
2017. Available from: https://www.
gzinnovation.eu/section/11/materials
[Accessed: Apr 11, 2018]

[19] Grado Zero Innovation. MuSkin—
The Mushroom Peel [Internet].
2017. Available from: https://www.
gzinnovation.eu/section/11/materials
[Accessed: Apr 11, 2018]

[20] Iaac Advanced Architecture Group.
Piel Vivo/Bio-plastica—Material
Explorations [Internet]. 2016.
Available from: http://www.iaacblog.
com/projects/piel-vivo-bio-plastica-
material-explorations/ [Accessed: Apr
11, 2018]

[21] Atelier LUMA, Studio Klarenbeek
& Dros. Labo Algues [Internet]. 2018.
Available from: https://atelier-luma.org/
projets/labo-algues [Accessed: Apr 11,
2018] (Photo credits: Antoine Rabb for
Atelier Luma)

[22] Wu Y, Shah DU, Liu C, Yu Z,
Liu J, Ren X, et al. Bioinspired
supramolecular fibers. Proceedings
of the National Academy of Sciences.
2017;**114**(31):8163-8168. DOI: 10.1073/
pnas.1705380114

[23] Tang S, Sandström A, Lundberg
P, Lanz T, Larsen C, van Reenen S,
et al. Design rules for light-emitting
electrochemical cells delivering
bright luminance at 27.5 percent
external quantum efficiency. Nature
Communications. 2017;**8**:1190. DOI:
10.1038/s41467-017-01339-0

[24] TechXplore. Japanese Researchers
Develop Ultrathin, Highly Elastic Skin
Display [Internet]. 2018. Available from:
https://techxplore.com/news/2018-02-
japanese-ultrathin-highly-elastic-skin.
html [Accessed: Apr 15, 2018]

[25] Shim BS, Chen W, Doty C, Xu C,
Kotov NA. Smart electronic yarns
and wearable fabrics for human
biomonitoring made by carbon
nanotube coating with polyelectrolytes.
Nano Letters. 2008;**8**(12):4151-4157.
DOI: 10.1021/nl801495p

[26] Kretzer M, Augustynowicz E,
Georgakopoulou S, Rossi D, Sixt S.
Shapeshift [Internet]. ETH Zurich:
CAAD; 2010. Available from: http://
materiability.com/portfolio/shapeshift/
[Accessed: Apr 10, 2018] (Photo credits:
Manuel Kretzer)

[27] Skipping Rocks Lab. Ooho! Water
You Can Eat [Internet]. 2018. Available
from: http://www.oohowater.com
[Accessed: Apr 15, 2018]

[28] Concrete Canvas [Internet].
2018. Available from: https://www.
concretecanvas.com [Accessed: Apr 15,
2018]

[29] Aaronwitz N. Whorl Console
[Internet]. 2017. Available from: http://
nealaronowitz.com/whorl-table/
[Accessed: Apr 15, 2018] (Photo credits:
Miroslav Trifonov (left) and Kerry
Davis Studio (right))

[30] Strozyk E. Wooden-Textiles
[Internet]. 2018. Available from: https://
www.elisastrozyk.com [Accessed: Apr
10, 2018]

[31] BMW USA. GINA Light Visionary
Model [Internet]. 2008. Available from:
http://www.bmwusa.com/Standard/
Content/AllBMWs/ConceptVehicles/
GINA/ [Accessed: Apr 10, 2016]

[32] Salvi R, Huygels B. The Moving
Mesh. Performative Folding Master

Course, Associate Professorship of Architectural Design and Building Envelope, TUM. 2016. (Photo credits: Sandra Persiani)

[33] Numen/For Use. Tape Paris. Palais de Tokyo/Inside, 20.10.14-11.01.15 [Internet]. 2015. Available from: http://www.numen.eu/installations/tape/paris/ [Accessed: Apr 16, 2018]

[34] Aoki J. Louis Vuitton Matsuya Ginza [Internet]. 2013. Available from: http://www.aokijun.com/en/works/louis-vuitton-matsuya-ginza/ [Accessed: Apr 16, 2018]

[35] Oxman N. Mushtari, Jupiter's Wonderer [Internet]. 2014. Available from: http://www.materialecology.com/projects/details/mushtari [Accessed: Apr 9, 2018]

[36] Beesley P. Hylozoic Ground. Liminal Responsive Architecture, Cambridge, Ontario: Riverside Architectural Press. 2010. ISBN: 9781926724027

[37] Hypermembrane. European Union's Seventh Framework Program managed by REA-Research Executive [Internet]. 2018. Available from: https://www.hypermembrane.net [Accessed: Apr 14, 2018]

[38] Brueckner A. Experiment Cyclebowl: A Pavilion of Cycles at Expo in Hannover. Basel: Birkhauser; 2002. ISBN: 978-3929638653

[39] Achim Menges Architect, ICD University of Stuttgart, Transsolar Climate Engineering Stuttgart. HygroScope—Meteorosensitive Morphology. Centre Pompidou, Paris. 2012 (Photo credits: ©ICD University of Stuttgart)

[40] Mungenast M, Tessin O, Blum V, Khuraskina O, Morroni L, Gutheil T. Fluid Morphology, 3D Printed Functional Integrated Buidling Envelope. Associate Professorship of Architectural Design and Building Envelope, TUM, support of Rodeca, Picco's 3D World, Delta Tower. 2017 (Photo credits: TUM Associate Professorship of Architectural Design and Building Envelope)

[41] ICD/ITKE Research Pavilion 2014-15: Interactive Panorama [Internet]. 2015. http://icd.uni-stuttgart.de/?p=12965 [Accessed: Apr 10, 2018]

[42] Oxman N, Keating S, Kayser M, Duro-Royo J, Gonzalez Uribe C D, Laucks J. MIT Medial Lab. Swarm Printing, Silkworm Silk Deposition [Internet]. 2013. Available from: http://matter.media.mit.edu/tools/details/Swarm-Printing [Accessed: Apr 9, 2018]

[43] Balmer A, Martin P. Synthetic Biology, Social and Ethical Challenges. An independent review commissioned by the Biotechnology and Biological Sciences Research Council (BBSRC). [Internet]. 2008. Available from: http://www.haseloff-lab.org/resources/SynBio_reports/0806_synthetic_biology.pdf [Accessed: Apr 16, 2018]

[44] Non GMO Project. Synthetic Biology [Internet] .2016. Available from: https://www.nongmoproject.org/high-risk/synthetic-biology/ [Accessed: Apr 16, 2018]

[45] Melchiorri J. Silk Leaf [Internet]. 2014. Available from: http://www.julianmelchiorri.com [Accessed: Apr 16, 2018] (Photo credits: Julian Melchiorri 2014)

[46] Blain M. Synthetic spider silk could be the biggest technological advance in clothing since nylon. Quartz [Internet]. 2016. Available from: https://qz.com/708298 [Accessed: Apr 15, 2018]

[47] Bolt Threads [Internet]. 2018. Available from: https://boltthreads.com/technology/ [Accessed: Apr 15, 2018]

[48] Shamees Aden [Internet]. 2018. Available from: http://shameesaden.com [Accessed: Apr 14, 2018]

[49] Callaway E. Glowing plants spark debate, critics irked over planned release of engineered organism. Nature News. 2013;**498**(7452):15-16. DOI: DOI 10.1038/498015a

[50] Krichevsky A, Meyers B, Vainstein A, Maliga P, Citovsky V. Autoluminescent Plants. PLoS One. 2010;**5**(11):e15461. DOI: 10.1371/journal.pone.0015461

[51] Synthetic Biology in Cambridge. Cambridge SynBio Startup Colorifix Wins Rainbow Seed Fund "Breaking New Ground" Award at Bio-start Competition [Internet]. 2017. Available from: https://www.synbio.cam.ac.uk/news [Accessed: Aug 11, 2018]

[52] Chieza N. Reflections from Ginkgo's First Creative-in-Residence, Assemblages 001, Ginkgo Bioworks [Internet]. 2017. Available from: https://www.ginkgobioworks.com [Accessed: Aug 11, 2018]

[53] Collins JJ, Maxon M, Ellington A, Fussenegger M, Weiss R, Sauro H. Synthetic biology: How best to build a cell. Nature News & Comment. 2014;**509**(7499):155-157. DOI: DOI 10.1038/509155a